MARIA BORELIUS

FINDING HAPPINESS AND HEALTH THROUGH AN ANTI-INFLAMMATORY LIFESTYLE

Wholeness • Food • Research • Exercise • Beauty • Insight

Translated by Sonia Wichmann

HARPER DESIGN
An Imprint of HarperCollins Publishers

FOR RITA AND ANNIE,
PATHFINDERS. SISTERS.

We shall not cease from exploration, and the end of all our exploring will be to arrive where we started and know the place for the first time.

—*T. S. Eliot*

CONTENTS

THIS IS A BOOK
ABOUT MY JOURNEY

At the age of fifty-two, I was experiencing menopause symptoms, back pain, fatigue, and a general feeling of melancholy about my life. I felt that everything was beginning to go downhill.

But after just a few months with a new lifestyle, my life had changed. I was happier, stronger, and pain-free—what had happened?

Through a series of remarkable coincidences, I realized that I had stumbled on something completely new—anti-inflammatory food—that could cure and prevent illness and even mysteriously put the brakes on aging.

That's how my journey of knowledge began, a journey during which, filled with wonder, I began to research this new landscape and discovered that it encompassed much more than diet. The contours of a whole new lifestyle emerged, and the clues came to me from many different people, each one amazing in his or her own way.

I encountered, among others, an innovative and inspiring fitness model in Canada; a professor in Lund, Sweden, working at the very forefront of research; an Indian health spa with a punishing enema treatment; a prehistoric hominid in Addis Ababa by the name of Lucy; one of London's most visionary dermatologists; unusually long-lived members of a religious sect in the outskirts of Los Angeles; a Danish TV celebrity who turned out to be fifteen years older than I thought; a gut group in the English countryside; a top geneticist at Karolinska Institutet, Sweden with a weakness for riddles; a hip-swinging yoga instructor at a New Age meeting in California; and an ethereal detective searching for human wonder.

Each one of them has played a role in this drama.

And so has my own experimentation. I have poked around among omega-3 fatty acids, probiotics, gluten, lactose, meditation, bone broth, Ayurvedic nose diagnosis, HIIT training, yoga, sunsets, inflammation markers, and spirituality apps. I have set out filled with curiosity but have often encountered failures and had to find a different path.

Step by step, I've felt my way forward as I worked to solve the puzzle of how low-degree systemic inflammation causes illness and what we can do to make ourselves stronger, happier, and healthier.

All of this has resulted in a five-point program to bring out the best version of all of us, a program that combines everything I've learned with the conviction that a lifestyle has to work on the practical level, in every-day life, and in our lives together with other people.

There's a health revolution happening right now, where a whole new way of thinking about food, exercise, rest, awe, and health is being constructed—and through this process, we are discovering new tools for living in a lighter and stronger way.

This is my story, and I'm sharing it with you in the hope that you will find inspiration, healing, and your path to empowerment.

Maria Borelius
London, November 2017

The unexamined life is not worth living.

—Socrates

1. A NEW YEAR

It's a new year, 2013.

I'm fifty-two years old, and I'm feeling puffy and washed out. Christmas has brought way too much of everything: pickled herring, gingerbread, cheese sandwiches on raisin-studded Christmas bread, schnapps, toffee, and boxes of chocolates devoured at lightning speed. And on the heels of this holiday excess came a New Year's trip to the shores of Kenya, with cocktails at sunset and three-course dinners with wine in the velvety African night.

The trip back takes twenty four hours. When we get home and I have to carry my bag upstairs, I feel like I'm eighty years old, even though I've just spent a week in the sun. There's a dull ache in my lower back, and my joints hurt. I'm in the throes of perimenopause, and my period shows up fitfully, on its own schedule. My feet are sore and swollen.

And then there's my belly. Or my "muffin top," as the women's magazines like to call it: a jiggling roll that desperately wants to spill out over the waistband of my jeans. These days, every visit to a clothing store ends the same way. After admiring all the figure-hugging pieces, I'm drawn like a magnet to long tops that cover and disguise.

I also have constant little infections and keep coming down with colds and sore throats. An ongoing low-grade urinary tract infection has led to repeated courses of antibiotics, which make me feel tired and a little sick.

This is what it's like to start aging. *Sigh.*

I guess there's only one direction to go now, and that's downhill.

So thinks a melancholic part of me.

Another part of me snorts. "Don't be so pretentious. Be happy you're alive! You have healthy children and can work. *Get on with life.*"

Fair enough.

But a third side of me is looking for something more.

It's part of human nature to want to improve yourself. You don't always have to accept the cards that life deals you. We want to shape our own destiny. The questions burn in me—because it's more than just my back, my belly, and my infections.

Whatever happened to that strong and happy younger woman?

She may still be strong and happy, but there are longer stretches between the bright days. More and more often, I wake up feeling melancholy, or "blue," as people say. I feel blue all over . . . or gray.

I regret all the things that I didn't have time to do with the children when they were younger. I grieve for my dead father and brother and for my mother, who is ill. I become annoyed more easily when I run into problems at work, and I see obstacles as personal defeats, instead of seeing them as challenges that can be solved with creativity and willpower, the way I would have done in the past.

I make a mental checklist.

How is that life balance going?

My eating habits are okay, I think. After the binge-eating lifestyle of my teenage years, my eating habits have gradually become normal. I eat what I feel like eating, which mostly means home cooking with lots of vegetables and olive oil. When I feel like baking a chocolate cake or mixing vanilla ice cream with pralines and caramel sauce, I do it without reflecting too much about it. On a hungry evening, I can easily put away three pieces of toast with plenty of butter, cheese, and orange marmalade and then feel vaguely guilty; I don't know exactly why.

But my everyday food doesn't feel extreme by any means. I love tea, which I drink in large quantities, just like my mother and my English grandmother, but I've cut back on coffee because it gives me headaches and makes me feel edgy and then tired.

I like exercising, but it's a journey without any compass.

I'll find a few newspaper articles about a new kind of exercise program and follow it for a week or two. I do a little jogging when I have time and the weather allows it. Light weight lifting at the gym a few times a week; a little swimming; a yoga class. Everything's possible, but nothing has any real shape except for the walks with our beloved dog, Luna. I meditate. And I can still remember my own mantra. All in all, I'm not a wreck.

Still, it's as if gravity is pulling me downward. Life is weighing down my whole being.

I have an appointment with my gynecologist.

"I think I'm a little depressed," I tell him.

"No, you're going through menopause," he answers.

Is all of this just to be expected? Should I simply resign myself?

That's not in my nature.

Buddha supposedly said, "When the pupil is ready, the master will appear." In the Bible, Jesus says the same thing: "Seek and ye shall find." The idea that you can learn new things by setting out on a journey to find insight and knowledge is part of our spiritual tradition.

So that's exactly what I do.

On a business trip to the United States, I happen to see a book on display in an airport bookstore. It has a typically American title: *Your Best Body Now: Look and Feel Fabulous at Any Age the Eat-Clean Way.* The woman who graces the cover is not a twenty-five-year-old model but a woman my age who is glowing with health. She seems to welcome me.

Her name is Tosca Reno, and she writes about her journey toward better health in an intelligent and convincing way. She describes how, in her forties, as an overweight and depressed housewife who would binge on ice cream and peanut butter at night, she managed to escape her depressive lifestyle and embark on a journey of personal health.

I can relate completely to the part about ice cream and peanut butter. I begin following her blog.

Tosca makes smoothies, does weight training exercises, and eats lots of protein. But suddenly one day, the content of the blog changes, from pleasant tips about healthy living to grave tragedy. Tosca's husband has lung cancer and only a few days left to live. Part of me feels ashamed for following an American health blogger's story of her husband's death struggle, complete with pictures from his deathbed. They show the dying man greeting Arnold Schwarzenegger, apparently an old friend of his. Good for both of them—but it's embarrassing that I'm sitting here reading all this.

In spite of that, I'm hooked.

Tosca Reno writes about her husband's final hours in an open and sincere way that invites her readers in. After his death and funeral, she finds a personal trainer who is going to help her move past her grief. This trainer is a blond Canadian by the name of Rita Catolino. The two of them begin training for some kind of competition in which Tosca is planning to participate in memory of her dead husband.

What *is* this? I think to myself.

But at the same time—who am I to judge someone who has just lost a loved one?

Tosca and her personal trainer, Rita Catolino, start blogging together about health, work, love, and their inner life. When the trainer writes, it sparks something in me. This is about more than just lifting weights or running. This is about inner light.

Around this time, along with two other women, I've decided to start an aid organization that will support vulnerable immigrant women by helping them to start small businesses. We plan to empower them through education, moral support, and microloans, so that they can realize their dreams of having work and income of their own. We're going to call it the Ester Foundation, and we've been preparing the launch for two years. Now it's about to happen. But the work is non-profit, and I have to squeeze it in between my regular work as an entrepreneur and journalist and my family responsibilities.

The paradox I'm facing is this: I will need *more* energy, but I have *less*. I think of the airline flight attendants and their oxygen masks. What is it they always say before the plane takes off? *Put on your own oxygen mask first, before assisting others.* I'm forced to lift myself up, *energize myself* somehow in order to be able to give to others and to carry out this project that I am passionate about. And the situation is urgent.

I suddenly have an idea. I'll seek out this Rita Catolino and ask her if she could train me too—online, across the Atlantic.

I soon realize that Rita Catolino is a kind of fitness star in a world that's foreign to me, where she trains women who participate in American bodybuilding and fitness competitions. Way out of my league, in other words.

So I write her an email.

Dear Rita Catolino,

I'm writing to you from across the Atlantic. I'm far from being an American fitness star; in fact I'm a fifty-two-year-old woman with four children and a heavy workload. In addition to my work, I'm about to start up an aid organization to support marginalized immigrant women. But if I'm going to have the energy to support others, I need to be strong myself.

That's why I need your help. I'm flabby, I have backaches, and I'm going through perimenopause. But I'm dreaming of something else. I need a plan.

Can you help me?

Best regards,
Maria B

Click.

Very quickly, I get a reply. She asks me to answer a number of questions and send pictures of myself in my underwear, and then we'll see.

My husband wonders where these pictures are going to end up. I tell him that they really aren't much to look at, and I send along both photos and questionnaire. And we—Rita and I—agree to work together for three months.

Then I receive the first training program. At least I think that it's a training program, but it's also about food, gratitude, and wholeness.

In many ways it's totally bewildering.

But three months later, my life is transformed. My body has changed shape, my muffin belly has melted down to its previous shape. And above all: my aching back has calmed down, and my inner light has grown brighter. I wake up feeling energetic and happy, full of faith, just as I used to earlier in my life. I feel stronger than I have in twenty years.

I get questions about why my skin looks smoother, what kind of exercise I'm doing, and what I've done to get a slimmer waist. People come up to me and tell me that I'm looking younger and happier. My brighter inner light somehow seems to have become magnetic. New and more positive people come into my life, with new ideas and a more positive flow. I also find a way to let go and to resolve a conflict that I've had with a close relative, which has been gnawing at me for more than a decade.

At the same time, I'm motivated to try to understand, on a deeper level, what is happening in my body and soul. Is there a medical explanation? Propelled by chance coincidences and a large dose of curiosity, I soon find myself on the very front line of medical research. It's about how low-degree inflammation affects the body and ages it prematurely. It's about a new body of knowledge that demonstrates the connection between inflammation and many of our common diseases. And it's about how an anti-inflammatory lifestyle, which is exactly what I had unknowingly embarked on, can counteract aging and decline, making you a stronger, smarter, and more toned version of yourself.

I will be making this journey on several planes.

First, geographically. I wish that I could say it was like in *Eat, Pray, Love*, Elizabeth Gilbert's astonishing story about how she traveled for a year to Italy, India, and Indonesia to find herself. But that's not what my life looks like. There's a job to do, a family to care for, bills to pay, extensive commitments, clients to serve, columns to write; in short, the million small obligations of daily life.

My journey will continue for four years, in small steps, while the rest of my life continues in parallel. Whenever I travel, for business or pleasure, I try to fit in a piece of what gradually becomes not only my lifestyle but also my passion.

The process will develop into a life story—about the enormous challenge of changing my lifestyle, about my many failures, but also about my slow and unexpected victories.

It will also be a journey of knowledge in which, using my background as a science journalist and biologist, I set out to examine facts from a range of different medical disciplines—puzzle pieces that I get from nutritionists, physiologists, geneticists, and psychologists. It's a journey right down to our human roots, and its goal is to find out why the anti-inflammatory lifestyle has changed my life and whether it could change the lives of others as well.

This journey of knowledge will not be the way I first imagined it at all. It will take me to completely unexpected places and force me to think about conventional Western medicine, which has so much to offer yet also needs to broaden its approach and become more open to the role of emotions, the whole human being, and the ancient traditions of wisdom and healing arts.

But above all, it will be a story about the growing health revolution that is happening here and now and is only just beginning.

What if I fall?

Oh, but my darling

what if you fly?

—*Erin Hanson*

2. MY BODY JOURNEY

The year was 1982, and Jane Fonda was sweeping across the world in yet another incarnation.

Like some kind of three-stage rocket, she had transformed herself from the space traveler Barbarella, by way of the Vietnam protests, to glowing fitness queen. Leg lifts and legwarmers were the order of the day, along with something called a "workout."

In Sweden, the fitness club "Friskis & Svettis" (roughly, "Health and Sweat") had attracted huge numbers of Swedes who just wanted to get some everyday exercise. With all due respect for founder Johan Holmsäter and his cheerful troops of exercisers, this was not my tribe. I never really clicked with all the big gymnasiums, the big T-shirts, and the loose shorts that might let everything hang out.

But Jane Fonda . . . There was something about her combination of glamor and discipline that spoke not only to me but to masses of young women that spring.

Jane Fonda's Original Workout.

The book had a cover that I still remember in detail. Jane Fonda with Farrah Fawcett–style hair, fluffily blow-dried back from the sides of her face. She's wearing a red and black striped leotard, black tights, and leg-warmers. Resting her left hip and elbow on the floor, she holds on to her right leg with her right hand, lifting it high, straight up toward the ceiling, while her left leg reaches up toward the right one. She looks happy and strong.

I bought her book and gave it a ceremonial place on its own shelf in my little studio in a run-down building in the Kungsholmen district of

Stockholm, where I was living directly under some heavily speeded-up amphetamine addicts. Their scruffy German shepherd barked every time someone came or went, which seemed to be around the clock.

At the time, my then-boyfriend had just broken up with me. The eternal theme: getting dumped, with the pain and humiliation that followed. Since he couldn't explain why he wanted to leave in a way that I understood, my natural interpretation of the situation was that I was lacking somehow; I wasn't attractive enough, smart enough, or good enough. My pain expressed itself in the form of binge eating. One day, I would have only cottage cheese and broccoli; the next day, large amounts of ice cream, cookies, and self-loathing.

And so it rolled along, in a cycle that alternated between half starvation and gorging on carbohydrates. I felt bad and often had headaches because of my chaotic eating habits. This affected my studies and my part-time job as a medical assistant at a nearby hospital. The apartment where I was living was cold, and I was forced to heat it by using the kitchen stove, turning on the oven and leaving the door open. It smelled like gas everywhere.

I had friends who would regularly induce vomiting. But I wasn't able to vomit on command—I was a failed bulimic. My weight could fluctuate by as much as ten to twelve pounds in a month. And when I ate extra, I punished myself by only drinking water the following day.

My friends and I tried all the diet methods that the women's magazines published, week in and week out. The Stewardess Diet. The Egg Diet. The Scarsdale Diet. A friend recommended the new Wine Diet, which was based on white wine and eggs, even for breakfast.

Jane Fonda's classic workout book and video came out in 1981.

"It's great, you don't even feel how hungry you are," she said.

But Jane had also suffered from food issues, which she had solved with exercise. She wrote:

"Go for the burn! Sweat! . . . No distractions. Center yourself. This is your time! . . . Your goal should be to take your body and make it as healthy, strong, flexible and well-proportioned as you can!"

These felt like powerful mantras for a woman who had just been dumped, a chance to find my way back through hard work.

I lay on the rug on the floor of my studio apartment and tried to imitate the pictures in the book. I had to move the little coffee table in front of my love seat in order to have enough room to do all these new exercises. I had the gas turned on in the oven as usual and the oven door wide open to warm up the apartment. It was noisy in the apartment upstairs as people came and went and the German shepherd barked.

It was hard to lift my butt 250 times, as Jane recommended, but the harder it was, the stronger was my feeling of rebirth. I would move through this pain, to something new and better. I wanted to be like Jane Fonda on the book's cover.

At around the same time, a good friend of mine was also dumped by her boyfriend. The two of us formed a self-help group for dumped women and spent several weeks dissecting our breakups and who had actually said what to whom. But our conversations always came to the same conclusion, a unanimous condemnation of two completely oblivious young men in Stockholm. Our judgment was broad, covering personality, morals, and looks.

After a while, my wise friend thought we should get off the couch, widen our repertoire, and maybe get a little exercise. And as I mentioned, by that time Jane Fonda had arrived in Sweden. It was a big event in what was then a calmer and more peaceful Sweden than the Sweden of today.

The newspapers *Expressen* and *Aftonbladet* reported on the worldwide fad that had landed in Stockholm, via a woman named Yvonne Lin.

Yvonne Lin was then world master in the martial art of Wushu, which I had never heard of. She had gone to Hollywood to learn from Jane Fonda and to absorb her training methods. In an underground training center on Markvardsgatan, a little side street off Sveavägen, Yvonne Lin started Sweden's first workout center.

Now we were going to try Jane Fonda for real.

We stepped into the studio as if into a temple, reverent and quiet—and immediately felt bewildered. A group of grown men were running around in the space, directed by someone who looked a lot like Bruce Lee, the martial arts master from Hong Kong. Instead of legwarmers, they had wooden pistols and were pretending to shoot at each other. One of them was yelling "bang!" as he hit a brick with a series of karate chops. I recognized two very well-known men who were often featured in gossip magazines. But where was Jane?

It turned out that the space was also used by Yvonne Lin's husband, who was a martial arts master, and that this was some kind of self-defense training.

We entered cautiously into the training studio. When Yvonne Lin stepped in, wearing a tight outfit with perfectly rolled legwarmers, and put on Human League singing "Don't You Want Me" with the bass pumped up, I was swept away.

This was completely new.

The workouts had the rhythms and choreographic awareness of dance routines. They focused on exactly those body parts that I wanted to re-shape; they had glamor, elegance, and humor and alternated between precision and free expression. There was an upbeat feeling to the work-outs, and they boosted our self-confidence, since we all worked in front of a large mirror, looking at ourselves for forty-five minutes. It was like being on Broadway, or participating in a lineup of dancers in *Fame*, where we would collectively dance our way to success and the perfect body.

Now, more than thirty years later, I can see the narcissism in this. The fixation on the body, disguised as neo-feminism, partnered with a business mind-set masquerading as health movement. I also remember Jane Fonda's almost desperately clenched jaw when I got to interview her on TV a few years later. She was a slim woman who seemed slightly fearful to me then—a far cry from the liberated workout rebel we had all believed in.

But she was a child of her time. The United States and Europe had left the hippies, unisex styles, and political demonstrations of the 1960s and 1970s behind, in favor of white wine and shrimp, Wall Street, padded shoulders, yuppies, and a new interpretation of what it meant to be a man or a woman. And yes, it was largely about the body and material things. Or as Melanie Griffith famously told Harrison Ford in the movie *Working Girl*: "I have a head for business and a body for sin. Is there anything wrong with that?"

The ideal probably lay somewhere in between. But we should look at our past with compassion and realize that maybe we needed a daily dose of Jane Fonda in order to grow up and become "whole" human beings. In any case, our little self-help group, "The Exes," needed a daily fix. And little by little, the feeling of being dumped faded away.

Gradually, the swinging food pendulum calmed down as well. I had a breakthrough one morning. I was sitting at the dining table at home in my apartment. The table faced out over a courtyard where two little children were playing. The night before, I had eaten sandwiches, ice cream, and candy. I felt anxious and guilty and was now considering whether I had the right to eat breakfast.

I drew a diagram, looked at it, and tried to think about what my relationship to food looked like and what feelings it triggered. Out of these thoughts an image emerged, a circle or spiral where crash dieting was followed by hunger, which was followed by overeating, which in turn was followed by feeling bad, which in turn made me feel that I had to start dieting again. It kept turning, around and around and around. Dieting—hunger—overeating—bad feelings—dieting—hunger . . .

I couldn't control the hunger that appeared when I had been eating only little broccoli florets and some cottage cheese for several days in a row. It was also impossible for me to control the overeating once it started. Nor could I control the anguish that overeating brought with it. But between the anguish and the decision to start dieting there was actually a little window—a window of willpower.

There and then, at the dining table, the thought struck me. I could feel anguish—but still decide that I was allowed to have breakfast.

A new spiral was born. It was a better spiral, where I always allowed myself to eat, even if I had overeaten the night before. Since I didn't diet as strictly anymore, I was less hungry and my indulgences became more modest, eventually tapering off. Jane Fonda gave me this victory. But it was a brittle harmony. I had to exercise in order for the balance to work.

Yvonne Lin decided to train workout instructors. We were a large group of hopeful young women who came to the audition that preceded the training itself. I was now a completely different person than I had been just a few months before. My relationship with food was more balanced; I was stronger and had higher and more consistent energy levels. And I was dependent on exercising, which had saved me.

When the audition came, it felt like a matter of life and death. I stood in a row with the other women and did aerobics like crazy. Even though I had never been much of an athlete, I hoped to be able to become an instructor, to be able to get into the training.

And I was chosen. When we gathered for the first time and introduced ourselves, all of Sweden was there. We were a cross-section of the country, cutting across educational levels and family backgrounds. We waited tables, we fixed teeth, and we worked in shops. We were students. We danced or taught. We were ordinary girls but also girls with mysterious occupations who seemed to glide around in Stockholm's underground/ fashion/artistic/glamor world. We formed a true sisterhood in our way-too-cramped dressing rooms.

When one of the sisterhood had just had a baby, her boyfriend cheated on her with a TV celebrity. After our training buddy found someone else's black lace undies in bed when she came home with her newborn—and when the TV celebrity also gave an interview in a tabloid where she talked about how she seduced men in carpenter pants—there was no end to the sisterhood and the primal power that came roaring out of our group. Wasn't the TV celebrity a snake and the boyfriend a swine? We watched over the abandoned mother like lionesses. No one would be able to hurt her.

We exercised for hours at a time, day after day.

And now I began to see the structure behind the training. How you started with a warmup, and then worked the shoulders, back, abs and waist, legs, butt, and finally abs again. There was a system. I also understood which types of exercises were good for each body part. And how to find your place in the music and count the eights correctly, with the beginning impetus of an exercise on beats one, three, five, and so on.

We learned how to stand, move, and speak in front of a large group of people and get everyone to move in the same direction—literally. How to get the energy and joy going and build up the participants' motivation. It was extremely useful.

We also learned to do things many times. Since we didn't use any weights, we added extra resistance to the movements and did endless repetitions—for example, lifting your leg 155 times at a certain angle. It required toughness, but we learned to be tough. That too was extremely useful.

I had studied physics and math in Stockholm, then biology. Biology was exciting and I wanted to continue, so when there weren't any courses in human biology in Stockholm that spring, I went to Lund. It was March when I came down from Stockholm by train, and the Lund night was damp, raw and cold. There were no rolling suitcases back then, so I was carrying two heavy suitcases from the Central Station to the apartment that a friend had let me borrow. The apartment was supposed to be furnished. That was debatable, as it turned out.

There was a kitchen table, a built-in bed, a stuffed eagle, and a saltwater aquarium with fish from a Norwegian fjord that the owner had caught during a course in marine biology.

At first I felt lonely in a city full of young people who all seemed to know each other. My genetics course had few students and didn't really provide a context where I could meet other people. And there wasn't anything like Jane Fonda's workouts or Yvonne Lin.

A thought struck me, and I called my self-help friend.

"We should open up something here," I said.

"Do you really think people are ready for it?" she asked.

I went looking for exercise spaces at a time when working out and gyms barely existed in Skåne, and I had to try to explain the concept when I met with landlords. We finally found a ballet studio near the All Saints Church. We would open our place there, a simple business with a big idea: to become the first Jane Fonda studio in Skåne.

I had another hidden motive as well. If only I could work out, I would be able to keep my eating in check.

A few years later, I had finished my education as a science journalist and had a child. Lund had not only offered opportunities to study and work out— I also met an incredibly wonderful man, and we fell in love and got married.

Soon I was expecting my second child. I was now working on the editorial team of an independent TV channel in Stockholm, a workplace with a fast tempo and lots of creative tension around a brilliant but tough boss.

Some women just develop an adorable little baby bump when they are pregnant. I've never looked like that. My belly was big, my legs were heavy, and there were still four months left until the birth.

Then I woke up one morning unable to walk. My lower back was incredibly painful and my legs wouldn't carry me. My husband drove us to the maternity center and had to support me as I walked in.

"You have a loosening of the pelvic ligaments," the midwife told me.

She gave me a pair of crutches. They helped a bit, and I shuffled out of there.

I felt like I was seventy-five years old as I limped in to work with my crutches, next to my young and childless co-workers. I had to swing one leg in front of the other in order to get over the threshold and down the stairs. Our tough but brilliant boss had a reputation for bullying people, and one of his former colleagues had advised me to always stand when I talked to him so as not to give him the upper hand. So when I spoke with him I would stand up and lean on my crutches, but I didn't feel particularly tough in all the struggles we had over how to do things.

My midwife associated the pelvic loosening with the physical and psychological struggle of communicating with my boss. It was caused by stress as much as by my body.

Things got complicated in the grocery store, as I juggled shopping bags and crutches, and was barely able to lift my hungry two-year-old.

One of my workout friends, who also was a naprapath, came to my home and looked at my back. She gave me some exercises that helped.

"Your ligaments are worn out," she said.

"What can I do about it?" I asked.

"You have to make sure you keep your muscles strong, to compensate. Never stop working out."

My eating habits were more balanced by this time. It was the early 1990s, and we ate a lot of pasta and bread, as people did in those days.

I gave birth to four children within five years and also had a miscarriage and an ectopic pregnancy that led to major surgery. After that, my lower back was worn out. The large central abdominal muscle, or rectus abdominus, had been torn in the middle, and I had scars from various complications. My female body had been subjected to the rigors of birthing and ground down by everyday life, but it had also been loved and nursed babies and was beginning to understand how wonderful life was. I was no longer a carefree young woman whose thoughts centered on men and studies. I was a mother with great challenges on the job and in the family.

It wore on my body. But I still felt strong.

Along with the children came an interest in food. In the past, I had struggled to normalize and find some kind of balance, but having to take care of the children transformed me.

In the early 2000s, my husband's workplace moved to Great Britain and our whole family followed. I began working from there, also in a new role, and became aware of organic food. It was a different country, where eating habits were completely different from the meatballs, quick-cooking macaroni, and fish sticks that had been our everyday fare in Sweden.

The grocery stores were bulging with processed junk food, and the results were visible everywhere. In the children's new schools, we saw a lot of overweight students, who stood around eating candy after school or sat in the schoolyard with a bag of chips. At the same time, there was a selection of organic fruits and vegetables that I had never seen in Sweden, where organic products in the early 2000s consisted mainly of small, wilted carrots.

Here the organic produce was greener and fresher. It was exciting. A new friend inspired me to begin making more food from scratch. She taught me how to make casseroles and showed me the Jewish chicken soup that she had learned from her mother-in-law that was better than penicillin. It clicked. Something in all of this reminded me of my mother's food. It was *real* food, the kind I had grown up with, the kind of homemade food that I used to eat, before single life, fast food, and stress messed everything up.

I found an article about the powerful effects of omega-3 oil and experimented with myself and my family. The oil seemed to make everything better: PMS, stress, anxiety, concentration problems . . . What kind of miracle oil was this? How did it work?

In an American magazine article, I found an interview with an American dermatologist with perfectly smooth skin, Dr. Nicholas Perricone. He talked about salmon as a miracle food that helped counteract wrinkles, stress, and anxiety. He also talked about something that he called "low-grade inflammation," as well as about food and disease prevention. I put the information in my fleeting internal memory.

Gradually, our family's eating habits began to change. We ate more homemade and organic food. We ate lots of vegetables, good fish, and poultry. Our butcher was situated in the English countryside, in an old shop from the nineteenth century on a winding country road, and also sold homemade applesauce and little jars of pickles that were lined up above the chicken breasts and roasts.

They also proudly displayed sausages that had won both gold and silver in the British sausage contests, hitherto completely unknown to me. These gold and silver sausages were made of real meat, from locally raised animals, and contained mixtures of lamb and mint or pork and leek. They were a taste sensation and became a staple food in our home.

I enjoyed baking, using good ingredients. Chocolate cake on Sunday with extra butter, berries, and cream. I no longer dieted. We got a dog, and walking the dog became my new workout, aside from some sporadic visits to a nearby gym. These were sunny years. Good years, shimmering years with a wonderful flock of growing children. Nothing could hurt us.

At least that's how it felt then.

Life's blows come in different shapes.

Some people go through devastating divorces. Others have children with serious illnesses. People are injured in car accidents or become ill with incurable cancer. You lose your job, go bankrupt, or experience other tragedies. You can feel as if your life has ended. For my part, the tsunami washed over me in October 2006—at least it felt like a tsunami at the time.

I was asked to go into politics. Not that I was a typical "partisan"; I had never really understood how you could see people as enemies just because their opinions were different from yours. It felt more like a kind of visionary military duty, to work on a number of issues that I felt were important, like research and entrepreneurship.

I was an outsider who made my way into a system that was hard to understand, and both the preliminary party election and the parliamentary election went unexpectedly well. Just in time for the 2006 election, I moved home from Great Britain with three of the children, while my husband remained with one son for a transitional period. I was elected to parliament and also quite unexpectedly became trade minister. The whole thing was unthinkably strange. But I had a dull feeling in my stomach.

After only a few days, a storm arose when I said that my family had paid a nanny under the table in the 1990s, long before my political involvement and before Sweden implemented the "RUT" tax deductions for household help. With four small children and my own business, as well as two ailing parents, I couldn't have made my life work any other way. Of course it was completely wrong—I realized that. But it was hard to explain myself once the machinery was set in motion. What I said in explanation sounded crazy or confused when it was printed. As an outsider in the political system, I felt completely helpless. I didn't have good political networks; I had no one to talk to and little support.

At home, the Swedish Security Service, or Säpo, explained that my family had received death threats and that they couldn't protect us since we didn't have any fence around our house. My children cried. We couldn't go out and walk the dog because there were so many journalists standing in the yard. We were on the front page of every newspaper.

Finally, I couldn't handle it any longer. I asked the prime minister to be excused from my post because I felt that I would never be able to perform any meaningful work at all. We were in total crisis, near a breakdown.

This is not the book in which I'm going to describe this in detail—the enormous lessons that I learned from being a non-politician in the political power center, about the powers and counterforces that arise, about the tough political game. And about myself and my weaknesses but also my unexpected fighting spirit and my great toughness. Perhaps I'll write about this someday.

In any case, the dramatic journey came to affect my inner life and my body—big time, as the Americans say.

We moved back to Great Britain, to my husband and the son who had stayed. I couldn't sleep for weeks, in spite of strong sleeping pills; I woke up every night in a sea of sweat and pinched myself in the arm.

Is it true that all this happened to me?

I was confused and shocked. Family members went into depression. I felt a deep sense of guilt for everything I had exposed them to but had a hard time providing the support that I wanted to since I barely had enough energy for myself.

Then I found Emelie. This ethereal woman was a personal trainer at a gym in the area, and she carefully trained me twice a week. When she massaged my back at the end of one session, my tears began to flow.

"Why are you crying?" she asked.

"Something terrible happened," I explained. "In another country."

She looked at me with her kind eyes.

"That doesn't mean anything right now."

But of course it did. The questions gnawed at me. Would anyone ever want to have anything to do with me again? My husband, who had never even felt I should become a politician, was fantastic in supporting all of us and bringing us back together. But I needed to find my inner strength again.

With her exercise sessions, Emelie helped me do it. My self-confidence began in my body, like a steady flow from her wonderful sessions. I strained and worked with my body and began to realize that I had been barely breathing for the last two months, just panting like a panic-stricken dog.

I also began having new thoughts that I had never had in my life before. I had experienced difficult times before, but they had always been about someone other than me. Now I saw things with new eyes. I thought about women's vulnerability, life's fragility. How could I use what I had learned in order to help others?

I looked up a well-known business leader in London who was on the board of a growing microfinance organization with extensive activity in India. At the end of the meeting he asked me if I would like to go there and see how I could contribute, and within two weeks I was on a plane to Chennai.

I ended up among some of the world's poorest women and children. The children crept up in my lap and gave me eager hugs. The women lent me their children across borders of skin color, language, religion, culture—I was incredibly thankful for that. My heart couldn't defend itself. They just crept right in, and I decided that I would process what I had experienced and turn it into light, for other people. It could begin here, with these people.

After a while I became CEO of the organization in London. The world was my field of work, and I gained many insights into life and fates far beyond what I could have imagined. It gave me completely new perspectives, a completely new sense of humility.

During this time, I learned a vast amount about our complex world. I was able to do hard things, big things, and work with exceptional people from all backgrounds.

I met poor and vulnerable women in India, South Africa, and Kenya and got to see the female power that helped give them the energy to start businesses to earn money for food and clothing . . . similar women, although with different skin colors, all over the world.

One day in Swaziland, the little mountain kingdom that lies in the blue haze of the southeastern corner of South Africa, I stood in front of a self-help women's group where all—yes, all—of the women showed traces of abuse. It was so common in the village that no one reacted to a black eye, or even a broken arm. The women came with bowed heads to the self-help group that we supported, and they left with backs that were a little straighter than before. I didn't even have words in my vocabulary to describe the struggle in their lives, the sorrow for those who became infected with HIV when their men had returned from working in the mines of South Africa.

It was huge and mind opening to see all this. One day I was talking to donors at the world's largest banks, and the next day I would meet with the world's most vulnerable people. I got to see everything—all the great and wonderful things, all the fighting spirit but also the vulnerability and awfulness. All in the same week. I learned an incredible amount and gained perspective, and things fell into place.

But it took a hard toll on my body—all these constant long trips that were often taken in the middle of the night, on a plane to or from Asia or Africa, as the only woman and sometimes the only European. I visited airports in cities that I barely knew existed just a few years before.

On a midnight flight between Chennai and Doha, I met Indian guest workers who were on their way to Qatar to build roads and football stadiums. One man told me that they were treated almost like cattle and worked under extremely hard conditions. Several of his comrades had died in workplace accidents. Their eyes were desperate, their bodies sunken. I will never forget that night.

In this context it felt a little shameful to think about my own body, so I stopped thinking about it. I didn't have time to think about it either, and with irregular meals and sporadic exercise, life began to wear me down. But just like with those oxygen masks—if you don't take care of yourself, you can't help anyone else either.

My first back strain came just like that, after three weeks of travel. I couldn't get out of bed for three days. A few years later, I had constant back pain. I walked around with little pillows to tuck behind my back when I sat and wrote. There were little wedge-shaped pillows in my bag, a manifestation of my new old-lady life. Not that I had anything against old ladies—just the opposite. But I was only fifty-two, after all. What would the rest of my life be like?

And exercise? It had dissipated, turned into an unengaged, unconstructed kind of activity.

"What was I going to do here?" I might ask myself when I arrived at the gym and drifted around randomly among the machines. A little cycling here, some weights there. It wasn't a catastrophe by any means. It just wasn't *me* anymore.

It was simply as if a gray fog had draped itself over my life. The children were getting older, and a couple of them had already moved away from home. It was empty. Who was I now, without children at home?

Sometimes the thought came to me that life would never be really sunny again. Was it menopause? Or was it that I couldn't move the way I used to anymore, now that my back had begun giving me trouble? The kids? I looked for explanations and had a hard time expressing what was missing. I just had a general feeling of malaise and depression.

That's how my life was starting to go.

And now we've arrived at New Year's, 2013. The moment of truth.

After the long trip home from Kenya, I can barely walk up the steep stairs in our house in London. I hoist the suitcase upstairs by swinging it, and my legs, in front of me step by step. This is the last straw. I lie down on the floor and put my legs up against the wall. Something has to be done. I send an emergency signal up to the higher powers and ask them to show me the way. It doesn't take long for the answer to come, in my own head.

"Why don't you get in touch with that woman named Rita, who trained the blogger Tosca Reno?"

I Google Rita Catolino and find a number of pictures. Rita is, let me just say it, a blond beauty with wonderful blue eyes, an open smile, and an incredibly well-trained body. What strikes me most of all is that she's glowing with health and strength. She has thousands of followers on social media. I myself have neither Facebook nor Instagram. It feels like a stretch for me to contact her.

A few years earlier, I had heard a good metaphor for inner dialogues—that inside every person is a struggle between two completely different beings. Or more specifically, it is the same being but different parts of the brain that are activated. One is the ape inside us, or the old parts, from an evolutionary standpoint, that lie in the center of the brain. The ape is governed by basic reflexes. We react to threats, stay with the flock, and take care of our offspring. We act on instinct, and catastrophe is always nearby. The other being, who acts inside us at the same time, is the human being, our higher self, which is guided by the frontal lobes, or outer parts of the brain. That's where those skills are located that human beings acquired later in their evolution. That's where we can use our good sense and plan ahead, but also interpret feelings in an empathetic way and withstand impulses that we know are confused or even dangerous for us.

My ape and my human being are now having a pretty heated inner dialogue.

"She's not going to want to take you on," says the ape.

"Why not?" says the human being.

"Because you aren't sharp enough. A hardworking career woman and mother with cellulite, fifty-two years old, doesn't belong in her fitness world."

"That's exactly why you need her," the human answers inside me. "She knows new things that you don't know yet."

"But it's expensive."

"What's the cost of having a ruined back?"

"What if she says no?"

"What if she says yes?"

Finally, I send my email. And I get an incredibly friendly answer. I have to complete a long questionnaire, and Rita also tells me to keep a journal of everything I eat for three days.

It's interesting to see what slips into my mouth during these days, especially one day when I have an early flight followed by a hard workday, and finish with a plane trip back in the evening. Hmm, let's see . . . olives, nuts, rye crackers, a piece of chocolate, a little bottle of wine . . . When I read through the food diary later I wonder if the airline had a single piece of food left on the plane when I got off.

But that's my life. I dutifully account for the three days, just as they were, and send off the answers to a number of other questions about old aches, exercise habits, energy, and sleep. I also have to indicate if I'm pregnant.

Um, I don't think so . . .

�֍

Then Rita's training packet arrives by email.

A new program for a new me.

It sounds promising and contains almost twenty different files that I open one by one, along with a message in which Rita promises to answer all my questions and asks me to communicate if I don't understand anything.

Let's see . . . Training . . . Hmm . . . It seems to be mostly about food. Is this a mistake?

I know about food already, and I eat well—I think. Except for certain exceptions, like that late night on the plane, but I had been working incredibly hard then, after all. I glance through the packet.

Eat homemade food. Less junk. More vegetables. Fewer trans fats. I know all this. *Old news.* Then we get to the order of the meals. Now there's some biochemistry. Certain meals should consist of protein, fruit, and fat. Other meals should only have protein and fat. A third type of meal should have proteins and complex carbohydrates. There are five to six meals every day with pure nutritional science. I understand the content, but what's the logic behind it?

Then it seems like there are certain foods you should eat. There are long lists of vegetables and allowable fruits. I see that bananas aren't included, a food that I eat every day. The only complex carbohydrates on the list are quinoa, sweet potatoes, and brown rice. And oats, "if you don't swell up."

I observe that there are foods that I already eat, more or less, but also foods that are new to me, like quinoa and chia seeds. And protein powder, which I don't know anything about. Most important, things that I really like are missing: crusty bread with butter and cheese; pasta; the occasional piece of cinnamon-topped apple pie, with creamy vanilla sauce; pickled herring . . . just to give a few examples.

So, I compose an email.

> *Dear Rita,*
>
> *Thank you for your tips. The exercise program sounds amazing. I'll do it. But the rest of it feels a little odd to me. I already have good eating habits and I like both bread and desserts. Why should I eat quinoa, but not pasta, for example? So, I'm following some of your advice but plan to do exactly as I like for the rest of it.*
>
> *Best regards,*
> *Maria*

No, that message doesn't get sent. And not the next one either, where I ask the questions I have about how everything fits together.

I can't quite explain why, except that I've simply decided to take care of myself. Partly I don't want to bother Rita, for some reason; partly I want to have space to do things my own way, which has been a small specialty of mine ever since my childhood.

I'll confess that at the beginning, I'm not completely on board. I decide to try a few little things now and then.

My first challenge is breakfast. How are you supposed to eat? For the past thirty years, ever since I cured my disastrous binge-eating lifestyle, I've eaten whole-grain bread, cheese, and eggs in the morning. Now I'm supposed to have warm water with lemon juice, pills, and a powder with a name that starts with "L." After that I have a few different breakfasts to choose from: protein powder with fruit, something called a "seed bowl," and pancakes made with coconut flour.

People are probably at their most habit-bound when it comes to breakfast, in particular, and these breakfast suggestions feel very foreign to me.

On the other hand, I dive in to the vegetables, fish, garlic, and olive oil with a feeling of both familiarity and happy expectation.

Then there's the exercise program. I realize now that this program is mainly about weight training, starting carefully and gradually increasing intensity. There are detailed instructions and references. For the first few days, I feel both uplifted and lost. I print out the program and make a little folder, then I sit down and Google the exercises to get the right balance and technique. YouTube turns out to be full of American muscle men who demonstrate in less than four minutes how to lift weights, while talking enough to give the expression "detailed description" a new meaning. I watch these videos when I don't understand something, then try it for myself. Above all, I'm buoyed by the feeling of having a plan at the gym. Most of it goes well, but some of the new exercises fill me with anxiety.

On one list is "dead lift." I Google my American muscle-building guides and see a man with a barbell on the ground in front of him. On the barbell are large round weights. He bends over and grips the bar with both hands and then lifts it up with straight legs and straight hips, with the bar hanging in his arms. He says that this is the Rolls Royce of exercises, with a gigantic effect on strength and back health, and that every fiber in the body becomes activated. I see how his whole back tautens and feel sheer terror.

How will I manage this?

I go to the gym to try it out, and I'm able to lift exactly four pounds in each hand, with bent knees. Then I feel a pulling in my back. When I look around, people are lifting 60, 80, or 100 pounds in the same exercise. Dead lifts are not my thing. Not at all my thing.

My first real setback comes a few days later.

I still don't understand why, but I develop an abscess in one armpit. It starts out as a small inflamed knot in a hair follicle, which grows into a golf ball at a dramatic pace. The thing looks grotesque, like a kind of

An anti-inflammatory snack.

baboon nose in the middle of my armpit, and is incredibly painful. I can't work out for a week. During this week, a car needs to be driven from Great Britain to Sweden, with a dog, and I sit in the car for twenty-four hours with my husband and the carsick dog in the backseat, elevating my arm by holding on to the handle above the door, while poor Luna throws up.

And so the first communication Rita has from me is not a well-written email with questions about why and how, but instead this:

Hi Rita,

I've come down with an abscess in my armpit the size of a golf ball, and have to sit with my arm elevated and can't work out. I'll be in touch when I feel better.

Maria

It sounds like the all-time worst excuse, kind of like "the dog ate my homework." But it's the truth.

The golf ball finally disappears, and I resume my new lifestyle. I move forward with baby steps and fall down all the time.

It's hard to follow the lifestyle at work. I'm out having lunch with a client, and I already know that she struggles with her weight. When she sees me order salad with smoked salmon and pass on the bread, she looks irritated.

"But you don't have to diet—look at me," she says.

"This isn't dieting," I say, defensively.

The intimacy that we used to have on a private level is marred by this conversation. I feel that she thinks I'm indirectly criticizing her, which in no way is true. I have friends who ask if I've become anorexic or developed a fear of fat when I turn down a piece of chocolate cake.

"Don't you eat anything anymore?" they ask.

"Yes, I eat lots, five times a day—I'm just eating different things."

Another friend accuses me of betraying the collective global feminism

A TYPICAL ANTI-INFLAMMATORY DAY

A typical day in my life might look like this:

- **6:30** Meditation and gratitude. Make my bliss plan for the day: food, exercise, de-stressing, awe.
- **7:00** Smoothie with protein powder, almond milk, green powder, spinach, berries, and nuts. Two cups of super strong tea with honey.
- **8:30** On my way to work, listen to my own bliss music.
- **10:00** At work I have two eggs, two rice cakes, and some tomatoes that I've brought from home, plus a cup of coffee with real milk.
- **12:00** Leg day at the gym—squats, dead lifts, hip lifts, etc. My bliss music in the headphones.
- **13:00** A protein shake and an apple. After showering, I eat a bag lunch with leftovers from yesterday (chicken/fish, potatoes, etc.) that I've added to a big salad with colorful vegetables.
- **17:00** A bowl of kefir with chia seeds.
- **18:00** Twenty minutes of meditation with my spirituality app or deep breathing.
- **19:30** Dinner—salmon fried in coconut oil and turmeric, oven-baked sweet potatoes, cooked green beans, homemade pesto, and a spinach salad, and then a few pieces of dark chocolate and a cup of ginger tea.
- **22:00** Digital detox—time to calm down my system for the night. Reading and gratitude list.

by focusing on my body and my food. I ask her if women will get higher pay just because I have back pain.

"But those are patriarchal ideals for women," she says, hurt.

"Is it feminism when women don't feel well?" I continue.

I begin to realize that anyone who starts a big lifestyle change will always have to deal with other people's reactions. Some of it is concern. Some of it is based on feelings. Suspicion? Anxiety about changes, because we want people around us to always be the same? Or does it come out of religion—a kind of asceticism, the idea that anyone who turns their focus on the body and their own lifestyle becomes self-absorbed?

I'm blown away by the resistance.

Rita and I begin communicating about all this.

I now understand that many people who change their dietary habits encounter exactly the same resistance from those around them—even at home. But Rita is not only smart and empathetic but also fun and ingenious, and she offers suggestions for how to meet these challenges.

She says that I need to stand up for myself and my lifestyle more clearly, without placing blame on anyone else. If others then choose to feel bad about my choices, it's *their own problem*. I need to learn this, again and again, and oh, how hard it is. I take it personally, and have always done so, if anyone in my circle feels bad because of something connected to me. I carry this like a heavy backpack, and I see the same phenomenon in many women around me. The trick is to lighten that backpack, since it's no use to anyone. Then there are the practical issues.

My family protests because the pantry and refrigerator are suddenly too full when I put in new, space-hogging things like bags of flaxseeds, hazelnuts, and goji berries. The freezer is packed with different kinds of frozen berries and big packages of frozen vegetables. My husband, who has many wonderful traits, has a strict inner home economics teacher—

we're talking sturdy cooking lady from the 1950s here. He loves a semi-fanatical order in the cupboards and the doors closed, which becomes hard to achieve when my new foods have to jostle for space with the foods we've always eaten.

And all these new powders, where can I store them? Like L-glutamine, as it turns out it's called, and green powders—a new phenomenon—and protein powder. That's also new, this thing with protein powder. I use it either as an ingredient in my breakfast, with nuts and fruit (protein, fruit, fat as it's called in Rita's language), or after working out. I find a kind of protein powder at my local health food store that tastes like banana muffins. The only problem is my stomach, which also turns into a banana muffin and starts to produce gas on a scale that could drive the heating system of a medium-sized town.

Another kind of powder turns my stomach into an even bigger balloon. Rita urges me to look for a protein powder that doesn't make me gassy, and she recommends a vegan powder that's easy on the stomach. That one is impossible to dissolve in water without a blender.

That's how I end up on a trip with a client to Geneva with my immersion blender packed in my bag. I arrive earlier at the hotel, and the first thing I do is go down to the gym and do the day's workout. Then I get out the wand from my luggage, and the powder I brought with me in a bag, and make a hotel room smoothie in the toothbrush glass, with the Swiss sparkling mineral water *Gerolsteiner Sprudel*.

In short, a *sprudel schmoothie*.

I've had better tasting drinks. But worse ones too.

Then there's my mood. Is it the spring light here in Geneva? My fun traveling companions? Or is it . . . me?

Something is starting to happen.

"My family protests because the pantry and refrigerator are suddenly too full when I put in new, space-hogging things like bags of flaxseeds, hazelnuts, and goji berries."

All my life through,
the new sights of Nature made
me rejoice like a child.

—Marie Curie, chemist and
Nobel Prize winner

3. INSIGHT

It's a spring night in Lund, Sweden, 2013.

It's just the kind of fresh spring evening that creates such expectations of life, love, and all the other wonderful things that belong to the light time of year. Students are riding their bikes toward the city center. Trees are budding in the Lundagård park next to the cathedral's sandstone walls. The magnolia by the cream-colored university building will soon begin to bloom, just in time for May Day, when student singers will once again sing a welcome to spring and the beautiful month of May.

Together with the other members of the advisory committee that meets regularly in order to support the university's big 350-year jubilee, I'm sitting in the old Biskopsgården, just below the library. At the last minute, I've decided to attend this meeting even though my calendar is full. It will turn out to be a significant event.

Every time the group gets together, we have the privilege of meeting one of the most innovative researchers at the biggest university in the Nordic region. Today we're going to meet a specialist in nutrition research. Professor Inger Björck is introduced and steps forward to talk about her brand-new research. Only a few minutes into her presentation, I realize that her findings are very important, even somewhat sensational.

She gives us a brief background.

Professor Björck leads the Center for Preventive Nutrition Research at Lund University. Scientists there are conducting interdisciplinary research about how a variety of diseases can be counteracted with a proper diet, as well as research into what is known as the metabolic syndrome.

The metabolic syndrome, a medical term that has become more and more common, includes three conditions: diabetes, obesity, and high blood pressure. Each of these conditions carries risks. But together, they form a type of super risk for serious heart disease, stroke, and other cardiovascular diseases. It is also suspected that this metabolic condition is connected to certain forms of cancer and even to an increased risk of dementia.

Researchers haven't quite been able to explain the metabolic syndrome. One theory is that it has to do with insulin, the hormone released by the pancreatic gland when you eat sugar-containing foods and whose function it is to move the broken-down sugar into the cells. People with diabetes 1, which often begins to manifest in the teenage years or even earlier, lack the ability to produce enough insulin.

But there is also an acquired form that sneaks up on people later in life, diabetes 2. (Today there are researchers looking into whether there may also be a number of intermediate forms between diabetes 1 and 2, but we'll leave that aside here for the sake of simplicity.)

To sketch out a simple explanatory diagram for this process, when you eat sugar and your blood sugar level rises, a signal is sent to the pancreatic gland, which releases insulin. The insulin is sluiced out and "opens up" the cells in order to sluice in the broken-down sugar, along with proteins and fat.

When the body constantly takes in large amounts of sugar and insulin levels have to stay elevated in order to shuttle the sugar out of the bloodstream and into the cells, it creates a so-called insulin resistance. In other words, there is insulin in the blood that's supposed to deal with the sugar, and that makes the insulin attach to the cells, but something goes wrong in the communication between the insulin and the cells. The cells simply lose their ability to react to the presence of insulin. The number of people who have metabolic syndrome is growing rapidly, because more and more people eat the wrong kind of food, have a sedentary lifestyle, and/or suffer from stress and other psychosocial problems.

The above-mentioned triple combination, with diabetes/belly fat/high blood pressure, used to be a medical condition that affected mainly older people. But now it's increasing even among younger men and women. Altogether it's estimated that one quarter of the adult population in the United States, Canada, and Europe have metabolic syndrome. In short, we are talking about an epidemic that is increasing like an avalanche in the Western world, an enormous threat to public health.

In the past, each of these diseases was studied separately. But Inger Björck and many other researchers worldwide are now beginning to realize that the diseases are in fact connected.

"Then you have to wonder, how can suffering be prevented?" she says.

Inger Björck is carrying out innovative research in this area. For example, she's studied mice that have been fed either a high-fat or a low-fat diet. In addition to that diet, the mice were given different berries and fruits like lingonberries, raspberries, prunes, and currants. It turned out that the mice who ate berries—especially lingonberries—maintained the same weight regardless of whether they ate a high-fat or low-fat diet. The lingonberry group actually lost some weight, even with a high-fat diet.

Björck believes that the risk of diabetes 2 and coronary artery disease can be decreased by means of an entirely new method, a new category of food in which berries are part of a larger food group.

"These foods are called anti-inflammatory," she says.

I make a note of the name. It calls to mind what I read about in Dr. Perricone's books ten years earlier.

Then Professor Björck begins to explain how these new foods can affect the whole person, not only blood pressure and cholesterol levels but also cognitive ability, or the brain function that includes a person's intelligence, in the broad sense of the word—our capacity to think, remember, solve problems, and learn new things. This research sounds both creative and worthwhile, and so far I'm following her presentation with interest. This is worth supporting, my professional self acknowledges in an observant yet slightly distant fashion.

But when she shows us the list of the foods the researchers have been using to achieve these results in people, I get a shock. A slow-motion lightning bolt strikes my brain, and I sit at the very edge of my chair, suddenly wide awake.

First, there are things like decreasing sugar, doing away with white flour, increasing the intake of all kinds of berries, increasing the amount of vegetables and fatty fish, and adding vinegar and probiotic supplements. But then comes a concrete list of foods, and it looks like . . . *Rita's food list*?

My heart does a quiet leap of recognition and time stands still. I gaze around me at the old meeting room, with its view of the university library's stepped gable in brick. The great linden trees shimmer with fresh new leaves in the spring evening.

What is this? Have I unknowingly been eating anti-inflammatory foods and thus affected my body much more deeply than I had realized?

The effects I've felt are exactly the ones that Inger Björck describes in her test subjects. They grew stronger, reduced their waistlines, expanded their mental capacity, and developed more of a zest for life.

Or is it just an amazing coincidence?

After the talk, we are served an anti-inflammatory buffet that the researchers have designed themselves. They've even baked their own bread, similar to Danish rye bread, using whole barley. There are salads, fatty fish, and nuts, and everything is delicious. Over one of the salads, I share my insight with another woman. I lean forward confidentially, almost a little embarrassed.

"I've actually been eating like this for a few months. Or at least trying to."

"I thought you looked more energetic, somehow," she says, looking at me appraisingly.

I go up to Professor Björck and tell her that there are in fact people who live like this every day but who haven't quite made the scientific connection to anti-inflammation that Björck's team has. They just do it because they've discovered that it works.

"Who are they?" she wonders.

"Well . . . fitness people in the United States and Canada," I say.

She looks surprised. We agree to stay in touch. And that's where my own journey of knowledge begins.

Inflammation and anti-inflammation. What is this all about? I have to learn more.

I begin racking my brain for long-ago facts from my university studies in immunology. I think I took that course in the red building at the old Veterinary College in Frescati in Stockholm, if I remember correctly, and we learned something about the two forms of inflammation—because inflammation is not always a bad thing.

The first type of inflammation is purely positive, a helping process. Imagine a cut from a kitchen knife, a finger squeezed in the car door, a urinary tract infection, or a sore throat. When you're injured or infected, your immune system starts producing inflammation as a defense mechanism. A teacher I once had used this image to describe it: Imagine a land that is being attacked by an external enemy and wants to defend itself. That's how the immune system works. The outer injury is the external enemy, the immune response is the country's government and defense, and the inflammation is part of what you have to do to defend yourself. There are a number of different foot soldiers who help. These soldiers in turn have many different specialist functions, just like in a regular army, with bridge builders, telegraph operators, explosives experts, and intelligence agents.

In human blood, the blood platelets constantly wander around looking for problems in the blood. The blood platelets gather around the problem—the cut, the bruise, or the infected body part—and then send a chemical signal to the immune system.

"Problem at g, come here right away," say the blood platelets.

The signal is intercepted by the white blood cells, who answer, "On our way."

An advanced line of defense is set up, with many different types of foot soldiers. They're called cytokines, leukotrienes, prostaglandins, chemokines, thromboxanes, and so forth, and they function like support troops, where each one sets out with its own task. They expand the blood vessels at the site of the affected tissue and make the area around it more "transparent." This means that more cells from the immune system can reach the injury, attack enemy bacteria, clean out old junk, and then repair and build up new and fresh tissue.

In medical training around the world and through the centuries, students have had to learn to recognize an inflammation the traditional way, which originates with the ancient Roman Celsus, who wrote great reference books about the body. Celsus's favorite treatment was to simply open the veins and empty out the "extra blood," a procedure he recommended for many types of health problems, as well as for people "with big heads." Celsus also described the signs of inflammation in Latin: *rubor, tumor, caldor, dolor*. Redness, swelling, warmth, pain. Which is exactly what you feel in your throat when you have a sore throat. These signs of inflammation can in turn be counteracted by RICE, or rest, ice, compression, elevation. (Exactly what you do with a sprained ankle.)

The whole point, in short, is that inflammation works like a kind of fire department. It rushes out, attacks the enemies, cleans out, and repairs. Then the system goes back to resting status.

This acute type of inflammation has a rhythm. There's an ebb and flow, a clear beginning and an end, and the rhythm signals a healthy and active immune defense. It isn't this type of inflammation that's problematic but rather another one, which seems to be affected by food and contributes to illness. Who might be able to tell me more about it?

I investigate some more, and after a while I find a new trail. There's a researcher in the United States, Barry Sears, who has been on this track for a long time and founded an organization for research in that area, the Inflammation Research Foundation. I'm not able to travel to meet him, but I don't want to just send him an email, since there's so much that I don't understand. We need to actually talk.

I'm able to reach him by phone, and he gets right to the point.

"This is a new area for most doctors. I've been working in the field for a while, but in general way too little research has been done."

He mentions how many different kinds of diseases the low-grade systemic type of inflammation is linked to. We're talking about heart disease, high cholesterol values, diabetes, joint problems, and neurodegenerative disease, but also certain forms of cancer.

"But what exactly does this low-grade systemic type of inflammation do?" I wonder.

He begins to explain very fast, and it's hard to follow him since the connection breaks several times during our call.

"Okay, how about this: I'll send you a scientific article," Dr. Sears says.

He soon emails me an article from *European Review of Medical and Pharmacological Sciences*. I click it open.

"The inflammatory response was developed over millions of years and allowed us to coexist with a number of microbes. The same inflammatory response also made it possible to repair physical damage . . ."

Okay, I think, acute inflammation is an ancient mechanism with benefits, millions of years old . . .

"But there are also equally important anti-inflammatory mechanisms in the inflammation cycle that allow cell repair and renewal. Only when these two phases are continually balanced can the cells effectively repair the small damages that arise with inflammation."

This is new to me. Does this mean that there's a need for balance inside the system itself—just as there's an inflammation yin, there also needs to be an inflammation yang?

"But if the proinflammatory phase continues in a low but chronic level under the pain threshold, it can drive many chronic illnesses. In the end it can result in organ damage, loss of organ function, and lead to severe illness, in spite of the fact that the initiating illness-causing events may have taken place decades earlier, triggered by an underlying and ongoing chronic inflammation process."

So, low-grade inflammation arises from imbalance—from a steadily ongoing inflammation that doesn't cause a "fire department" type of acute inflammatory response but in the long run can act as a catalyst for small seeds of illness that have been germinating in the body for a long time.

Is this the type of inflammation that we bring about through an unhealthy lifestyle? In other words, might bad nutrition, stress, environmental toxins, and other lifestyle factors give us inflammation, which in turn makes us sick? Perhaps that is why the wrong food can lead to illness and not just to us ingesting too many calories.

And is it true that long before we actually become ill, the low-grade inflammation affects us so that we start to "lose steam"? When I went to the doctor complaining about my back pain, depression, and listlessness and looked for explanations based on external things ("the kids are moving away from home"), maybe it was actually a low-grade inflammation, an imbalance in my immune defense caused by a number of lifestyle choices, leading to my bad back, blue mood, and bloated stomach. And maybe this is what I've "cured" with my new lifestyle choices?

I go on looking to see if my symptoms, like back pain, fatigue, and a "low" feeling, could have been signs of low-grade inflammation. I find the following symptom list:

- The skin looks older, is drier, and has more wrinkles
- Lower energy
- Less stamina when exercising
- Swelling in the face
- Swelling around the belly
- Increased risk of either constipation or loose bowels
- Less ability to concentrate
- Fluctuating appetite
- Fluctuating blood sugar levels
- Weaker immune defense
- Joint pain
- More depressed mood

I can tick off several of the points but not all. So far, we're just talking about what a doctor would call "everyday troubles." But how does inflammation work in relation to serious illnesses?

I realize that I'll have to become a detective in order to get to the bottom of this riddle. No single researcher seems to have the whole picture. I will have to solve a jigsaw puzzle.

�֍

A few years earlier, an editor in a publishing house gave me a book called *Anticancer*. I didn't read it then, but one day it falls off the bookshelf as if some friendly soul in there wants to help me on my way. It turns out to be a good lead.

The book is by the French neurologist and Doctors Without Borders activist David Servan-Schreiber, who developed a brain tumor at the age of thirty and set out on a journey of knowledge to save himself. In the

"I realize that I'll have to become a detective in order to get to the bottom of this riddle. No single researcher seems to have the whole picture. I will have to solve a jigsaw puzzle."

book, which became a bestseller in many countries, he reported on some of the leading research about the essence of cancer, as well as strategies for keeping up resistance. Servan-Schreiber eloquently describes how cancer and inflammation are intertwined and drive each other on in a kind of evil witch dance.

A tumor is a number of cells that begin to grow wildly and unchecked. In the beginning, there's enough nourishment for the tumor in its immediate surroundings, but after a while it outgrows its small neighborhood. The tumor now begins to operate with a devilish intelligence, causing an inflammation around itself. Why? Fascinated, I continue reading. The tumor uses the inflammation to manipulate the immune defense and make it "attack" the tumor from inside.

Once the immune defense has gotten into the tumor, it faithfully begins to work according to its usual procedure when it encounters inflammation, which involves, among other things, producing certain substances that are going to help repair the tissue. It's just that the tissue being repaired this time is an enemy—the tumor itself. The immune defense is literally fooled by the tumor. Instead of protecting the body against the tumor, it begins to fuel its further growth out into the body. New blood vessels are built to bring new nourishment to the tumor, and little supporting structures help to anchor the tumor even further.

To sum up, the tumor creates an inflammation that in turn feeds the tumor, which in turn creates even more inflammation in its surroundings, spreading the disease further. The effect of the inflammation is like pouring gasoline on the cancer fire. That's why cancer is such a diabolical disease and so hard to fight.

Professor Björck has also explained that inflammation is linked to coronary artery disease, obesity, diabetes 2, and joint problems. Is it true then that inflammation is either the basic cause, or least the promoter, of our main public health diseases—the diseases that cause so much human suffering—as well as aging and human breakdown?

And how does inflammation work in general? Is it like a wildfire that burns down the healthy parts of a human being? Or more like a flood wave that beats and beats against a barricade that finally falls apart? Or is it more like a low-level conflict between two people that distracts and weakens them so they are no longer able to defend themselves against an external threat?

Which one is the most reasonable scenario? I must keep searching.

But right now, I can state one thing that seems obvious: low-grade systemic inflammation is harmful and either triggers or speeds up disease. At this stage it's also apparent that there are foods that counteract the broad negative effects of inflammation and that these foods to some extent are similar to the Rita Diet, which is like the Inger Björck Diet—and also like the David Servan-Schreiber Diet, which kept him alive for almost twenty years after his brain tumor diagnosis, even though he was supposed to survive for only a few months.

I have found a lot to think about, and I'm encouraged about my new lifestyle. In general, I've started to like the "Rita program," as I still call it. And I've begun to feel results. They are modest results, but noticeable. My body is stronger, my belly flatter, and I'm sturdier both in my psyche and across my shoulders.

"You seem stronger, Mom," says my older daughter, unexpectedly.

That's good. I want to feel strong, and my new lifestyle grounds me with a new feeling of security. I'm slowly gaining more insights into this lifestyle, about what it is and what it's like to live it and not just talk about it. It's both surprisingly simple and complex, since it demands a new kind of awareness.

To have an anti-inflammatory lifestyle was never a goal in itself for me. I hadn't even heard of this as a lifestyle until that fresh spring evening in Lund, when I was already a few months into my new lifestyle. I just thought I would get a training program via the internet.

The fact is that I don't have time to spend dealing with food and exercise, I don't feel like losing weight, and I can't spend all my energy on it since I have a life to live too. You have to live your life in the human village, as Mowgli says in *The Jungle Book*. You can't live a life that's too different, because that's like settling down on a dry little patch of grass by yourself outside the village, surrounded by your pills, protein powders, and strange food. As a mother of four, I neither can nor want to live like that. After all, I live in a very loud and lively human village that consists of family, job, and friends, a context that's much bigger than just me.

But still I'm driven onward by this new feel-good sensation. The biggest change is that I have to start planning for eating well, to go from a lifestyle where I eat whatever I happen to find, or what tastes good, to strategically planning my food intake for health.

People say that if you fail at planning, then you plan to fail. Everyone who has children learns to plan food at home to some degree. It doesn't work to come home from work tired and have hungry kids digging through the refrigerator. (Those evenings always end with fries, fish sticks, and ice cream . . .) You just have to learn to be a few steps ahead. It's easy when it's about the children, but to think like that about my own nutritional needs is something I've never done.

The first thing I need to learn is how to eat in a more conscious and planned way, and that also includes thinking about my specific needs. It sounds pretentious and, above all, time-consuming. Let me explain.

We humans have a limited window from the time a feeling arises to when we want to act on it. The more we're aware of that window, the more impulse control we have and the smarter we get. But when it comes to food, hunger, and eating, this control is being disabled by the miraculous innovations of the modern food industry.

Today we can get hungry one minute and theoretically find food within the hour, as long as we don't find ourselves in a kayak on an expedition along the northeastern coast of Greenland, or looking for hidden treasure in inner Amazonia. There are little cookies in the pantry and fig marmalade in the fridge. At work, there are some leftover cookies by the coffeemaker. At the counter at the 7-Eleven are ready-made sandwiches. Our ability to plan food and think strategically about food doesn't bother trying anymore. It simply isn't needed.

I begin to think about how I in particular, and human beings in general, have ended up here.

Just imagine if we were as spontaneous about getting ourselves to work, for example. We would get up and get ready, and just as we were leaving the house, we would begin to think about how to get there and what address we're going to. But of course we don't do that.

Most people check the calendar in advance to see what time the meeting is, Google addresses, check that the car has gas, look up the metro lines, and see how far we have to go between the station and the meeting place. Not many of us would get to our jobs or our meetings on time if we didn't do all this. We need an inner map. A road plan.

We need this for food as well.

This is what I have to learn—that in the pause between feeling and action, there's a rainbow leading to a pot of gold, and it's easier to find that pot if I'm well prepared.

My basic plan becomes this: I plan how I'm going to eat as soon as I wake up in the morning. I plan for a good day. Many people do that anyway when it comes to work, family, and leisure activities. Why not do it for your own health as well?

In Rita's plan I wasn't given calories, quantities, or forbidden foods. Instead, I have a number of guidelines. The most important thing is to eat food that is as unprocessed as possible—food that you could pick, fish, or hunt. "Made by nature, not by man," as someone I met said.

Rita doesn't just want me to reduce sugar—something that I've known I should be doing for a long time—but also to avoid bread and pasta, which get broken down into glucose, or sugar. She wants me to replace these with sweet potato, quinoa, and brown rice. She wants me to eat protein-rich foods, often and in large quantities. Four or five times every day, I'm supposed to eat eggs, turkey, mussels, shrimp, fish, meat, or vegetarian protein. Can I even eat that much protein? I'm supposed to eat lots of leafy greens and vegetables, preferably four times a day. And good fats like olive oil, coconut oil, and nuts. All this advice goes into planning four or five meals per day.

Now this advice needs to be transformed into habits that will work in my everyday life. Then I have to have time for work and also exercise four times a week. It's stressful. How is that supposed to happen?

I can be undisciplined and lazy, with a tendency to overeat. Even worse, I tend to eat for emotional reasons: when I'm anxious, bored, or exhausted; or when I just have a craving for something good and make the usual mistake of satisfying this craving with food that ends up giving me only momentary relief.

How am I supposed to manage to eat in such a disciplined way?

I face several big challenges, which begin as soon as I wake up. I continue to look for a new standard breakfast. I don't want to have to think in the morning, when I'm a little sleepy and everything's spinning around in my head. What can I come up with?

Most of what goes into a typical Swedish or British breakfast is wrong, according to the new thinking. Juice, bread, yogurt, cheese, rolls, cereal—none of that works anymore. So I look for something that can become *the new breakfast.*

I test different things and arrive at smoothies for breakfast. Almond milk, berries, nuts, and protein powder. It breaks up our family's mornings, since my habits are so different.

Snacks are simple: a couple of hardboiled eggs and a tomato; nuts and fruit. But dinner demands more thought.

I was no cook before I became a mother, but once I had children I

ANTI-INFLAMMATORY VEGETABLES AND MUSHROOMS

Think of the rainbow—purple, blue, green, yellow, orange, and red. The more colors you eat every day, the prettier your plate and the more beautiful you will be, inside and out, since each color represents a certain kind of active polyphenol.

- Asparagus
- Beets
- Bok choy
- Broccoli
- Brussels sprouts
- Cabbage—white, red, cauliflower, green cabbage
- Celery—celery root and stalks
- Cucumber
- Dandelion leaves
- Eggplant
- Endive
- Fennel
- Kohlrabi
- Mushrooms—white mushrooms, pennybuns, oyster mushrooms, chanterelles
- Nasturtium
- Nettles
- Onion—red, yellow, garlic, leeks, spring onions
- Parsnips
- Peppers—red, orange, yellow, and green
- Radishes
- Salad—arugula, iceberg, mâche—go wild!
- Spinach
- Sprouts—alfalfa and all others
- Tomatoes
- Watercress
- Zucchini

Certain vegetables, like beets, parsnips, and celery root, have a higher glycemic index (GI) value than others. Mix them with vegetables that have a lower GI value, for example beets on a bed of arugula with a dressing of vinaigrette and nuts. Perfect!

became interested in cooking to nourish the family and create a happy mealtime. In my old life, it was easy to make food taste good and dress things up with extra butter, sugar, cheese, and breading, or by frying, adding good bread toasted with garlic butter, and so on. There were soup and pancakes on Thursdays. My husband cooks just as often, usually with extra everything.

I still want to eat good food, feel satisfied, and enjoy food together with my family, by myself, or with friends or colleagues, so I have to become more creative. But I don't have all the time in the world.

I decide to compromise. I plan meals with food that is natural but with a little glamorous twist. A little more taste, a little more spice, good sauces and dips made of tomatoes, avocado, grilled vegetables, spices, oils, and garlic.

The trick is to achieve good proportions. A plate divided into four parts, where 25 percent is protein, 25 percent salad, 25 percent other vegetables, and 25 percent rice or quinoa—more or less.

But there are many challenges.

"Where's dessert?" asks my son, with his big brown eyes. "You used to make that good chocolate cake."

It's true. Since I started cooking with my new method, I've increasingly lost interest in baking big, fluffy cakes. It's not about body weight but just the feeling that I want to serve my family something other than 2 cups of sugar, which my former prize cake contained.

So I experiment, with mixed results.

"Sorry, Mom, but this is a failure," my blue-eyed son laughs when I serve his best friend some zucchini cake.

The friend is too polite to say anything, but he stares listlessly at his piece of cake. A few strips of zucchini are swimming around like threads in the dry almond flour.

My brown-eyed son brings his new girlfriend home, and I serve them some protein muffins. I've found a recipe with protein powder, sweet potato, and almond flour. The new girlfriend smiles but doesn't take seconds.

My son grunts.

"What *is* this?"

It sounds like I have spoiled children, but I don't. They're just used to a different kind of food. It's said that Chinese children don't like cinnamon buns. Why? Because they never eat cinnamon buns. You like what you are used to. This way of eating is the opposite of how we used to eat, and the change takes time. But I don't really care; I have patience. I feel happy in some way. It's not just the spring light. It's something more—hard to put into words.

Then I find the explanation. Again, by chance.

I'm working on a book that I've been thinking about for a long time.

I once had a brother who died. My handsome, mischievous, idolized brother got sick in his twenties and was diagnosed with schizophrenia, a grim psychiatric diagnosis. In 1986, I lost him in a fire in a Stockholm apartment. Through a contractor project I've done for Karolinska Institutet, I've begun to think a lot about the stigmatization of mental illness.

Now I've decided to write a book that illuminates and looks into the taboo around mental health problems. This also involves dealing with the taboo within myself, the shame that I've felt—because mental health problems are looked at differently than physical disease. Aside from the sorrow, there's this damned feeling of shame that rests over both the afflicted and their loved ones. And that makes us doubly ashamed. We're ashamed because people we love have a shameful illness, and then we're ashamed because we're ashamed.

I root around eagerly in everything that's connected to this issue. I talk to researchers, read, and interview lots of people with different illnesses, as well as doctors and nurses.

While I'm looking through the latest research, a new branch emerges. It has a very long name: psychoneuroimmunology. It's the study of how mental illness can arise in the brain, and how it's linked to—here it is again—inflammation. Hmm . . .

In other words, on the one hand there's a connection between immune defense and inflammation, and on the other hand, a connection to brain health? Fascinated, I look more closely into this connection.

We've already mentioned all the foot soldiers that are sent out by the immune system. Among them are the cytokines, triggered by inflammation to show up in huge numbers—something called a cytokine storm. This storm, like a swarm of bees, starts up the body's defense system in the form of the so-called B and T lymphocytes. But the cytokines also talk directly to the brain.

Let's take that again. The immune system and the brain *talk to each other.*

This is a new piece of knowledge, a new puzzle piece. I investigate further.

The American researcher Robert Dantzer did the pioneering work that showed that the cytokines triggered by inflammation also affect the brain's signaling substances: dopamine, serotonin, and noradrenaline. Since these substances directly affect how we feel, physically and mentally, cytokines can change how we feel in emotional terms.

When you have a high inflammation level, the cytokines decrease the levels of dopamine, noradrenaline, and serotonin. You get a feeling of illness, like when you're coming down with something. You feel low, tired, withdrawn. And when the inflammation decreases, the number of cytokines also decreases, and the signaling substances can flow again at a normal level in the synapses of the brain.

I add this to what we now know about signaling substances, highly simplified. Balanced dopamine levels provide more energy and self-confidence. Balanced serotonin levels lead to more calm and less anxiety. Balanced noradrenaline levels lead to increased alertness.

That's exactly the change that I've felt in myself. This is interesting . . .

Not only does this train of thought offer new possibilities for understanding how mental illness begins, but perhaps it might also account for my new, brighter mood. A signal sent directly from my decreased inflammation level up to my brain might actually be affecting my mood. Has the new diet rearranged my brain chemistry?

I have to keep digging.

Researchers can demonstrate a connection between the degree of inflammation and depression, as well as between the degree of inflammation and the risk of suicide.

Suicide is today the most common cause of death among young men. One of the explanations is that there are too few resources available in the scandalously downsized psychiatric acute care centers. The doctors are forced to make a brutal selection among all the people who are seeking help, asking themselves terrible questions like "Who is actively likely to commit suicide? Who can we consider to be managing adequately at home, in spite of their depression?" They are forced to look for those patients who have the highest risk for suicide and send home the rest even if they are feeling unwell.

Since the price of making the wrong judgment call is so incredibly high, people have looked for more objective markers, something that can be measured, instead of simply asking the patient questions. As most people who have known someone who committed suicide realize, a person who really wants to commit suicide will hide it.

At Lund University, the researcher Lena Brundin found that in people with depression, the will to commit suicide was directly linked to the degree of inflammatory markers in the blood. Not only that, but the degree of violence used in the suicide could also be correlated with the degree of inflammation.

In the fall of 2017, new research was presented in London, where scientists from the University of Cambridge argued that there is a "very robust link between inflammation and depressive symptoms." Professor Ed Bullmore, chief of psychiatric staff, pointed to the fact that people who have just received vaccinations and people who take inflammatory medicines get depressed more often. The teams are now thinking of depression as a physical illness that might be treatable with anti-inflammatory measures.

It turns out that 30 percent of people who suffer from inflammatory diseases like rheumatism are also depressed, making that group four times more likely to develop depression than the general population.

Schizophrenia has also turned out to have connections to inflammation, in research carried out at the Karolinska University Hospital by the psychoneuroimmunologist Sophie Erhardt, a pioneering scientist I had the privilege of meeting when we both became involved in the Swedish Psychiatry Foundation's work. The same goes for bipolar illness.

It's clear that cytokines are linked to poorer mental health for people, and cytokines are produced when there is inflammation.

I'm now hearing more and more researchers say that there's a real connection between immune defense and the mind. Might these mental illnesses actually be immunological diseases? Which one is the chicken and which is the egg?

More and more doctors are coming to radical conclusions.

"Our old model of care, where we make a distinction between body and mind, is completely outdated, where psychiatric care is provided by psychiatric specialists and physical care by doctors and nurses who specialize in the body. We have to begin to educate people within healthcare who can bridge this gap—between immune defense and the nervous system," thunders Professor Robert Lechler, chairman of the British Academy of Medical Sciences, in an interview in the British paper *Daily Telegraph.*

Everything is connected, and the link is inflammation.

This is the very front line of research. I'm standing right at this front line and probing it as I'm writing the book, and I see the inflammation trail grow red hot again. I have to dig deeper, even though it's sometimes tough going—very tough.

I have the twenty-five-year-old grief of a big sister simmering away inside. It's been shut up in a closet with the door bolted shut and marked with a sign saying "Open at your own risk!" In that closet lives the grief I feel for not being able to save my brother. It sometimes feels like I've gone straight down into a black hole while I'm working on the book. I also encounter the sorrow and anxiety of the people I interview, people who have been stricken with serious illnesses and sometimes met with little understanding from the outside world; who feel alone and vulnerable even though they're fighting with such courage. It touches me at my very core, since I understand them all too well.

But then I notice something. The afflicted and their families say almost exactly the same thing: when they eat junk food, or bad food, their symptoms get worse. When they choose better food, the symptoms decrease.

The new lifestyle that I'm learning about shines so brightly in the midst of all this darkness, and it's signaling from all directions. It turns into a kind of elevator that leads me up toward joy, out of my gray mine shaft.

Up in the daylight again, a journey to completely ordinary things—things that might be trivial but that absolutely need to work, things that used to be self-evident before, in my old life, but that I now have to relearn.

Like how to shop for food, for example.

I used to wander around fairly randomly and pick out things that looked interesting when I wasn't shopping for a recipe or based on sale prices. I bought things mainly based on what my family likes to eat every day. Chips, bread, jam, cereal, milk, chicken, pasta, muffins, and vegetables. Nothing strange. That's what a regular shopping list might look like.

Now I'm starting to see the grocery store in a whole new way. It has its agenda, I have mine. That's why it's important to examine the grocery store's setup. You are often met by freshly baked bread that's meant to tempt you with its warm aroma, and then you're supposed to walk all the way inside the store to find the milk, a product that almost everyone buys. The vegetables are often hidden far inside, along some wall.

I decide to outsmart the store's selling agenda and my own old reflexes. I'll get a maximum amount of good and nutritious foods while minimizing gluten, lactose, and sugar, and I'll shop economically.

The first step is to make a plan for the day's meals every morning. Breakfast, lunch, dinner, and snacks. And then shop according to that. Just like an architect, you have to begin with a drawing in order to build a good house.

My plan might look like this:

Breakfast: Smoothie with protein powder, green spirulina powder, chia seeds, raisins, blueberries, and spinach.
Snack: Boiled egg, tomato.
Lunch: Chicken, sweet potato, raw grated carrot, and cooked broccoli.
Snack: Fruit and nuts.
Dinner: Lentil patties, spinach, and tomato salad.

If the kids are eating at home I add things that they like, but only then.

I'm beginning to dig around a lot more in the vegetable bins. I'm starting to pick up onions, tomatoes, carrots, lemons, garlic, broccoli, green beans, cauliflower, brussels sprouts, squash, eggplant, and so on, according to season and price; I inspect them and smell them. I find green cabbage. And white cabbage! This is an unassuming but wonderful, cheap delicacy—especially in the springtime, when the delicate spring cabbage arrives. Here I also find my clumsy, ugly, new best friend—the sweet potato.

I buy blueberries, especially if they're on sale, since you can freeze them. Strawberries and raspberries according to the season. Lots of frozen berries. Rita doesn't want me to eat too many bananas since they have a high GI value. Okay, we'll try.

I'm starting to think about the store in unpoetical terms. Like for example "protein shelves." That's where there are chicken fillets, meatloaf, pork chops. The egg shelf, and the shelves with canned sardines, mussels, and tuna, are also protein shelves. What has good quality and reasonable prices?

I often come home with different kinds of fish, preferably ethically sourced. Chicken thighs have more taste than breast fillets, and you can buy them in bigger packages with six or twelve thighs and then freeze the part you don't use in smaller bags. I buy according to season, price, and quality. Cans of mussels, salmon, and sardines, and quick protein solutions with lots of omega-3 fats. And also lots of eggs. They have to be from cage-free, happy chickens. I also buy beans and lentils of all kinds and shapes, since it turns out not everything is a good fit for my stomach.

I buy low-lactose milk, yogurt, and sometimes soy yogurt. I often try different kinds of nut milk, like almond, coconut, and hazelnut, and soy milk. I use butter once in a while, preferably organic.

The spice shelf expands. New tastes turn up there, and more experiments. At the base are of course salt and pepper of different kinds, and now also turmeric, which I'm beginning to learn is extremely anti-inflammatory. But other spices reduce inflammation as well. I check lists and find cinnamon, oregano, cumin, coriander, thyme, rosemary, basil, different kinds of chili, garlic, ginger, capers . . .

I buy different kinds of oil and begin flavoring it myself. A sprig of

ANTI-INFLAMMATORY SPICES

- Basil
- Capers
- Chili
- Cinnamon
- Cloves
- Coriander
- Cumin
- Garlic
- Ginger
- Lovage
- Oregano
- Rosemary
- Thyme
- Turmeric
- and many more!

rosemary, some garlic, and a few lemon peels quickly add a new taste in a couple of days. I try new kinds of vinegar—there are so many to choose from. I learn more about my trigger points—whipped cream and toasted bread.

I become a seed and nut eater and also buy lots of dried fruit, with favorites like goji berries, dried apricots, dried plums, figs, and cranberries. Little delicacies.

I put all these little things in plastic jars in a row at home.

My usually good-natured husband bangs around angrily among all the new jars that are crowding out his tubes of caviar, fig marmalade, and cheese, when he's in his home-economics-teacher mood. We start having new types of arguments. About foods in the cupboards. What goes where? It is not dignified but it is the new reality at home.

I also learn to make more food than I need.

Apparently, this is called "food prep" in bodybuilder language. You're prepping food when you grill long rows of chicken thighs, for example, and save them in the freezer. Or boil sixteen eggs at once. Or make a big batch of vegetable stew at a time.

Rita thinks I should cook in bulk twice a week so that there's always something at home that's easy to fix. I wonder if I have the time, but I soon discover that it doesn't take more time to make food in advance. It takes *exactly the same amount* of time, sometimes even less. But the difference is that you eat better when you've planned better.

But what if you're not eating at home? This will be a big challenge for me. With work in several countries and with children who are studying or working abroad as well, the year includes many days of travel. At such times, I'll set off early, on crowded morning flights where they serve sandwiches packed in plastic and a cup of coffee, and return late on other planes, where they serve even more sandwiches in plastic and more coffee. Food on the go, food in canteens, meals with clients—always on the road to somewhere.

How will I manage this?

It will be especially hard when I'm headed out on a really long trip to a completely different corner of the world, where I might be able to get a few more leads to how all the remarkable things I'm experiencing actually fit together.

❧

*Ayurveda is the holy science
of life and serves the whole
human being. Both in this life
and the next one.*

—Charaka,
the father of Indian medicine,
c. 300 BCE

4. HEAL

When you travel to India, the plane flies through the night and over the Indian Ocean.

On this night, there is so much turbulence that the red safety belt sign never turns off. No food or drinks can be served, and the toilets are closed for hours. Luckily, I've learned to bring food with me. Little cherry tomatoes, almonds, and protein bars become my salvation when the food cart is chained down all through the shaky trip.

We land in Mumbai early in the morning. I see an older woman in a sari leaning on her son's arm. She looks pale and worn out from the trip. We all desperately want to use the bathroom. But I'm continuing on and take the bus to the domestic terminal. It's been a few years since I was last here. The development has been rapid.

What was then like a sea of walking people now consists more and more of people on motor scooters, often carrying two or three people. Young women dressed in saris sit behind the men in their white shirts and black gabardine pants. The women sit sideways, sidesaddle, with a tight grip on the waist of the person in front; they travel at high speed on the road between terminals. Through the bus window I watch this bustling city pass by.

India has everything, extra everything, of everything.

More colors and more joy, but also more pain, and a vaguely menacing feeling. The poverty hits you like a blow to the gut. We pass slum districts where little children play among piles of garbage and puddles of brown water. But beyond the poverty, there are many other aspects to India.

India has one of the world's most sophisticated and cohesive systems of integrative medicine. It's called *Ayur-Veda* in Sanskrit, the ancient Indic language that was spoken by India's conquerors around 2000 BCE and is distantly related to all the languages of Europe, even our Nordic ones. *Veda* is basically the same as our word for wisdom, and *ayur* means life, youth, and health. When I was working in India, people explained this to me as the eternal and genuine knowing.

I've been invited to a course especially for women. It's a leadership course, but Ayurveda treatments are also included. Might I find more knowledge there?

I'm going to head to one of India's most advanced health spas for Ayurveda, outside Thiruvananthapuram, the capital city of the state of Kerala. The name of this city is almost unpronounceable. I just have to cross my fingers and hope that I've booked a flight to the right place.

When I arrive, I get to meet a doctor.

An Ayurvedic doctor is not like a medical school doctor, like the ones we are used to in Europe. She does take my blood pressure and measure my pulse. She is also professional, with a white jacket over an exquisite sari in blood red and gold. She asks me about any apparent diseases and ongoing medication, after having first asked about my medical history and past surgeries.

But that's where the similarities end.

I get a questionnaire with forty questions. What kind of food do I like the most? What kind of exercise do I do? Which smells and sounds do I react to? Digestive habits are extensively handled, as are sex drive, sleep rhythm, and the color and intensity of my dreams. The doctor looks through the questionnaire and makes some remarks to her assistant.

"Sweet, sour, salty, and bitter." She nods meaningfully.

The doctor creeps closer to me and suddenly peers up my nose at very close quarters. She listens to my voice and my way of talking. She takes my pulse for a long time and gently rotates my hand on my wrist.

Then the last question.

"You like bitter tastes?" she asks again.

"Yes, I do," I say, thinking of my favorites: tea, Campari, and arugula.

"*Vata-pitta*," she tells her assistant.

Both of them nod solemnly. A decoding of this follows.

Ayurveda describes the human being as consisting of three basic elements, or *doshas* in Sanskrit. *Vata* is the creative and innovative aspect of a person. *Pitta* is the organized and structured leader aspect. *Kapha* is the warm and integrating element in all of us. We have all three of these aspects in us, but in different amounts—partly based on our innate constitution but also varying by season, climate, and during different phases of life.

If you have stronger *vata-pitta*, like I do, it means that these elements are more prominent. It also implies that you more easily succumb to illnesses that are linked to these so-called *doshas* when in a condition of imbalance and stress.

For *vata*, it's the stomach that easily goes on strike, or the nerves. For *pitta*, it's often the skin or the heart and lungs that are sensitive, and too much *pitta* stress results in increased aggression and perfectionism. For *kapha*, it can be weight gain, lethargy, or depression.

A simple and quick test of which *dosha* is dominant is to imagine that you're sitting in a traffic jam on the way to a meeting, and you're late. The traffic is moving at a snail's pace.

If you become anxious and think the client you're going to meet will cancel your contract, the *vatta* element is dominant.

If you get angry, think that all the people around you are driving stupidly and are idiots, and you start giving the finger to people in other cars, you are a *pitta.*

Do you sit there calmly listening to the radio and figure that there's no point in getting stressed out? You're the *kapha* type.

Another categorization is by basic body type. *Vattas* are naturally slim and have a hard time building up muscle, *pitta* is the natural athlete, and *kapha* is the one who gains weight the most easily and builds muscle mass.

Ayurveda is a life art that describes both health and sickness, my doctor tells me. It was born from the observations made by tens of thousands of Ayurvedic doctors over thousands of years, all over the former India, which used to consist of a number of independent kingdoms governed by a long series of maharajas and nawabs. All of these hundreds of thousands of observations about how the body and soul of human beings worked were put together into a larger system.

The interesting thing about Ayurveda is that disease is described as something that takes place in several systems simultaneously, when too much total stress gathers. (Again, the idea that stress triggers inflammation.) I wonder if this might be the first system in the world that actually describes how low-degree systemic inflammation affects people's health. Perhaps a person who is seriously studying human health intuitively gets a feeling for inflammation and anti-inflammation?

That's why I'm curious about Ayurveda, and during this week in sunny Kerala, my plan is not just to attend a leadership course but also to carry out some private studies and to investigate this ancient healing art more deeply.

When we in the West reach the stage where we begin to treat an illness, Ayurveda considers that it's already too late. Illness must be met at the gate, early, before it's had time to develop into a full-blown disease, by actively counteracting the stressors that make disease develop. But there are many differences between the two ways of looking at health.

"The greatest difference is that we see that people are different. You in the West think that all people should have the same type of treatment," my Ayurvedic doctor tells me.

Instead of standard treatments based on the same criteria for everyone, and standard doses, they believe in individual treatment based on the needs that a person manifests through their *vata*, *pitta*, or *kapha* type.

Could Ayurveda possibly be describing the very thing that modern medicine is now beginning to investigate: the idea that low-grade inflammation can lie simmering in the body and contribute to making a variety of diseases surface—and that the specific illness is determined by the type of innate vulnerability the individual has? Today medical researchers are beginning to talk about the need for individually tailored treatments, and all doctors know that people respond differently to different medicines. But Ayurveda has always held that point of view.

"We also believe that food is medicine," says the doctor.

"How do you know that?"

"From observation," she says. "If you look at enough people and see the same things over and over again, you can see a pattern."

In Ayurveda, food is even considered to be the most important medication, more important than everything else, simply because we humans eat so often and take in numerous nutrients through our food. These nutritional elements, my doctor says, have the full capacity to either build, protect, and heal the body or create stress and disease.

This way of looking at the connection between food and health in many ways resembles the conclusions that conventional medical science is now beginning to draw. The difference is that most doctors who are educated in academic medicine and practice in Europe and the United States barely talk about this with their patients.

At the end of my first Ayurvedic medical visit, I receive my personal treatment schedule.

06:00	walk
07:15	meditation
09:00–12:00	treatment
16:00	meditation

In addition, we'll be attending the course for several hours each day, and we're given a number of different tasks to do during the week of the course, in groups or individually; we also need time to try to illuminate deeper sides of ourselves, or what our course leaders call our "shadow sides," to find out how they affect our ability to work. All of this in one week.

That's a lot to do, I think as I move into my simple bungalow, which is situated close to other little bungalows. You might describe them as an Indian kind of little cabin, in the middle of a semi-jungle of vegetation with abundant, large green leaves. For some reason, there are three ominous ravens watching over my patio. There is a plastic table and two aluminum chairs, and along the side, a laundry line sways in the wind. The ravens come flapping in with their powerful beaks as soon as I've had breakfast the next day. They eat everything, even the paper label of the tea bag. I see a few monkeys climbing around a little farther away.

Here wild herbs grow everywhere you go, because the health resort grows all the plants that are used in treatment and food preparation. The next day I look at the elegant handwritten signs that have been stuck into the ground, to see if anything looks familiar. *Abutilon indicum*? Some kind of mallow-like herb? *Acacia catechu* seems to grow into a mighty tree. There's a thin shrub with stubby little leaves that I don't recognize at all. I ask a passing doctor in a white coat about the plant.

"Ah, that's a *guggulu*. Very good for hemorrhoids."

People are sitting in a long line on the veranda of the treatment house, waiting. Women, men, Indians, Europeans, Asians, old and young. Doctors in white coats bustle around a large table on which a bunch of papers with scrawled notes are spread out. It's time for the female head doctor to assign the therapists who work at the center to those of us who have just arrived.

There are rows of Indian therapists sitting on the veranda. Almost all of them are women, dressed in yellow treatment clothes, a tunic and pants—sweet, soft women with gentle smiles. The only therapist who's standing is a completely different type of person. She stands with her feet widely planted and arms crossed over her chest. She has an eagle nose, a Clint Eastwood gaze—though with brown eyes—and she looks like she would take no prisoners. I hope I don't get her, I think sulkily, like a schoolchild who doesn't want to end up with the strict teacher.

The doctor begins to assign therapists, and it takes time. I tune out, in what I now understand is classical *vata* manner, captivated by all the beautiful herbs that lie neatly arranged in a display case in the entryway. Cinnamon bark, dried berries of some kind, powdered green herbs, pulverized ginger and turmeric . . . hmm, anti-inflammatory, all of them . . . star anise, cloves, black pepper.

Finally it's just me and the brown-eyed Clint Eastwood left. Of course.

The doctor points.

"You'll take her," she says.

So it's me and Shaila, it turns out, who will spend three hours a day together for an entire week.

It's going to be intimate.

The Ayurvedic patient isn't allowed to wear any clothes, except for a pair of little paper pants that are too small for me—I can barely pull them over my thighs. Shaila knows a few words of English. She gives me short commands.

Undress! Sit! Be still!

Then she begins.

Shaila massages the entire surface of my head, my shoulders, my breasts, my abdomen, my butt, every piece of cellulite I ever have or have had, every inch of my knees, thighs, calves, the inside and outside of my toes and fingers—in short, every last little molecule of skin that I have.

The first day, I get a whole-body massage. The next day, I get to bathe in quarts of heated oil. On the third day, Shaila pours a thin, even stream of oil on my forehead. All with a smooth, herb-scented oil that apparently is customized just for me. My body, lying on a ritually carved wooden bench, gets kneaded, massaged, stroked, thumped, folded, and slapped. The bench is made from a single tree trunk that's been hollowed out in the middle so that I can become one with the bench, they tell me. In the beginning, I feel incredibly embarrassed by the intimacy of this close contact with Shaila, under a very bright light. But then I begin to relax. And enjoy. I fall asleep again and again. It feels delectable to bathe in this oil. Shaila is professional and turns out to be very kind.

After having learned a little more about Ayurveda I can now venture to diagnose my therapist. I open my eyes under a stream of oil and ask:

"Are you a *pitta*?"

"Yes, me *pitta* lady," she says, and smiles for the first time.

After that she calls herself "*pitta* lady." "Shaila *pitta* lady," she says, massaging me with eyes that have become a little softer. Shaila and I become one. And as with all people, there are more levels than you would guess.

She has a fantastic sense of humor, which you need to have if you're an Ayurveda therapist, given what you have to see and put up with in this intimate struggle with oils, injuries, and decrepitude. She is also strong, persevering, and very smart. And she likes tips. When it's time for a tip, she gets the "Clint" look again and examines the money closely, which is completely understandable considering that she is probably supporting several people on her salary.

In the dining hall we can choose our food from a large buffet that is mainly vegetarian but also includes some small pots of chicken. Next to each serving dish there's a sign: Vata, Pitta, or Kapha.

There are wonderful vegetable broths and stews made of lentils and root vegetables. There is very little of what I call "substantial proteins" in the buffet, meaning animal proteins, and I have to supplement with some of the protein bars I've brought with me and extra fried eggs that I've ordered, or buy extra fish from the fish seller who comes onto the property. Otherwise it's vegetarian curries, *daal*, and vegetables fried in oil. And rice in all its forms.

I soon observe that it's hard to be a *kapha* since their food is low calorie with very few carbohydrates. It seems that *kapha* mostly get to eat steamed vegetables and a little rice. The *vata* and *pitta* food is richer; there are more thick sauces and vegetables in oil and garlic. It's good, even if it's a bit one-sided. But we are always hungry and eat a lot.

We women who are attending the course go to lunch straight from our treatments. I'm sure I've never seen such a dreadful-looking group of people, myself included. Based on what we've told the doctor about our everyday aches and pains, combined with our *doshas*, she has prescribed various therapies for us. As a result, my classmates and I turn up for lunch transformed beyond recognition.

We have yellow herbal masks on our faces, with black rings for our eyes. We have oil treatments with little grains of ash in our hair and hard paper turbans on top of that. We have plastic cones on our heads. Little twisted paper cones on our heads. Piles of one thing or another on our heads. The oil drips around us, leaving big oily stains on our poorly fitting green treatment jackets. Frankly, we look like crap.

In short, we look like we belong in a Monty Python film about an Indian health resort.

Ayurveda emphasizes the importance of the correct daily rhythm as a way to decrease stress. A good twenty-four-hour rhythm, in Ayurvedic terms, entails turning out the lights at ten and getting up with the sun.

At around six in the morning, we gather in the warm darkness, at the gate to the spa. We walk briskly along the narrow, heavily scented streets in the little town that is just waking up. Indian disco music jingles out of an invisible loudspeaker as the sun slowly rises.

When we turn down toward the sea, the sun is beginning to climb in the sky. We continue, walking briskly through a village. A sleepy older man comes out of his little house and stretches his arms up toward the sky. Women rinse themselves outside with rubber hoses. Kerala's fishermen twist their white *dhotis* around their waist, like Gandhi, prepared to set out on the rumbling sea. Scary, hungry, stray dogs run after us in the streets, barking.

When we arrive, we drink warm water with lemon, ginger, and honey. It tastes wonderful. Then it's meditation for those who wish.

And then yoga, which I only truly discover now. We do yoga with a teacher who takes us through the poses with stable, calm energy. The patio is sleepy. A fan swishes. Swoosh . . . Swoosh . . .

We have all the time in the world for this yoga practice, in contrast to the yoga lessons in Stockholm or London or Los Angeles, where I've felt an undercurrent of big-city stress and competitiveness. There's nothing like that here. In one corner, there's a fully dressed lady who does a little of whatever she feels like, wiggling a few toes. In another stands a man who can barely bend his knees. A few Europeans are working hard on their routines, as we like to do, and take every Downward Dog and Cobra to its extreme.

I let myself experiment with not trying to achieve anything in particular. Can you do yoga like this—just breathing calmly, feeling your way in every pose, sort of gliding into the pose? Yes, it works. The teacher walks around, watching us. He looks like a slightly overweight Swedish civil servant, but his hands are soft when he twists my hips.

I go along with everything that the doctor and Shaila plan for me. I use up a ton of brown towels, the towels we're given specifically to mop up all the oil. I take pills for this and for that, sprinkle dried herbs on my food, and drink tea in more shades of brown and green than I knew existed. I open myself completely.

Except when we get to something that I begin to hear people whispering about.

Panchakarma.

This word is whispered reverently at first, almost ominously. Apparently, this is something extra remarkable. When my fellow classmates have been prescribed *panchakarma*, they come to breakfast pale and trembling. I eventually realize that they've spent the night on the toilet, since the night before they were prescribed the most devilish laxatives that this mild little health spa can muster.

"Every hour," one sister from Sweden states laconically. "All night."

When it's my turn, I put my foot down and demand facts about what suddenly seems like moonshine.

"*Panchakarma* is not my thing."

"But it's important for the body to cleanse," says the doctor.

"Cleanse what?"

"Eliminate your toxins."

"What toxins?"

"Everyone has toxins."

"Prove to me that I have toxins in my body."

"But it's an important part of the cure."

"I refuse."

"You won't have the full effect otherwise."

"Sorry, but I'll only do it if you can show me that I really am poisoned. Otherwise, no thanks."

We are locked in a conflict. And that's how I leave it. The doctor is an empathetic person who retreats when she realizes that the battle over *panchakarma* is lost as far as this particular patient is concerned.

But this little episode by no means dims the overall impression.

I begin to feel a sense of total balance and a kind of superwoman

Maria—stronger, healthier, and more harmonious.

energy. Along with the deeper insights that the course gives me—when my body's relaxation is combined with an open mind and a deeper reflection about how all of us, myself included, spend so much time judging others through the glasses of our own reality—this becomes a totally transformative journey.

Open, calm, happy. I recognize the feeling.

De-inflamed?

Intuitively I feel that the cumulative effect of the massage, the food, the calm movement, the meditation, and the herbs I've been given is anti-inflammatory. Could these be the very things that are at the top of the anti-inflammatory list?

On my last evening at the health spa, I run into one of the younger doctors. I ask him if it could be possible that Ayurveda counteracts inflammation effects—that the treatments are in fact anti-inflammatory.

"It's possible. Many of the herbs we use are indeed anti-inflammatory."

"But what do all the treatments do?"

"We know that the treatments together reduce stress, which in turn reduces inflammation. But we don't express it the way you do."

"How do you express it?"

"We talk about a system in balance; we say that the whole is working," he says.

I want to know *why* the system is balanced, but he remains vague, to my way of thinking.

As I mentioned, Ayurveda is a wisdom tradition that builds on three thousand years of experience. It's an empirically based tradition, which is also being studied, but within a kind of traditional, wholeness-focused framework, instead of through the type of scientific studies that European and American research within conventional medicine often builds on.

Ayurveda simply hasn't fitted itself into the boxes of modern Western medicine. That's why the likable young doctor can't give me the answers I'm looking for, since he bases his answers on the figures of thought of his own tradition.

Since I have grown up in a Western thought tradition and been educated in its scientific ways of thinking, I'm looking for a direct explanation, one that says, "This is where it happens, right *here* something goes wrong, becomes better, or changes." Which way is best? The idea of wholeness that is often associated with Asian medicinal branches like Ayurveda and traditional Chinese medicine? Or Western medicine?

Well, Western medicine is often mocked for its squareness. But the squareness is also a blessing, since it helps us to weed out quacks, half-truths, and humbug. It does so by demanding straight facts in order to be able to establish a new idea as truth. These facts are delivered with the help of the entire arsenal of control systems that we've been developing, with some difficulty, ever since modern science was born in ancient Greece.

This control system includes clinical studies, with trial subjects who undergo new treatments that are tested under controlled circumstances. They include control groups where other, similar people participate in the study at the same time without receiving treatment so that the results can be compared, preferably ensuring that the trial subjects don't know whether they're receiving medicine or sugar pills. The controls also include peer reviews, where the results and methodology of the studies are evaluated by other scientists before they're published and allowed to become a new truth. The control system also includes the requirement of reproducibility. That is, other researchers should be able to obtain the same type of results if they perform the same experiment in the same way. Therefore, scientific articles are written with great exactness, so that others will be able to imitate the studies and check the results.

Thus, there are a lot of sentries along the way, from the point when a certain idea arises until we begin to believe that this idea actually is valid, and it becomes truth. This is the type of truth that is completely fundamental if we're going to base our health-care systems and treatments on facts and, in the bigger picture, build up some kind of reasonable society.

I have a deep respect for the three-thousand-year-old Indian system, its intelligent observations and apparently accurate conclusions, as well as all the wise people within the framework of this system, who knead and massage and think. But it's not enough. I don't understand why no one has tried to do scientific—or as I see it, "real"—research on the Ayurvedic ideas about herbs and potions in order to verify, for example, whether *panchakarma* might after all be the best thing that's happened to mankind since sliced bread.

But before anything like that turns up, I have to continue searching.

My demanding brain, which has been trained to think critically when confronted with new facts, wants clinical studies, control groups, peer review—all of that.

With that very brain, I want to understand if food really can be medicinal.

On lingon-red tussocks
and on shifting sands
Where pines are soughing,
susilull, susilo . . .

*—"Flickorna i Småland,"
a Swedish song by Karl Williams
and Fridolf Lundberg*

5. SALMON AND LINGONBERRIES

On the theme "there's no place like home," I'm back in Sweden. Here I will soon get to hear the story of a stunned doctor in Dalby and forty-four typical, everyday people who became radically healthier in just four weeks by eating anti-inflammatory food.

But let's take the story from the beginning.

It's a gray day, and a thick fog covers the ground. I feel that little rush of joy inside, as I always do when I'm driving toward Lund and see the big dome that rises above the plain in the distance. There is the side street where my father grew up, and I remember a June night outside the Freemasons lodge behind the apse of the church, my young self in a gold skirt that's way too short and a first kiss with a special person. Lund of the alleyways, Lund of student life, Lund of the cobblestone streets. But also, the growing Lund, the city of science that stretches beyond the highway and right into the future.

The Center for Preventive Food Research is situated at the outskirts of Lund University, right next to the main office of one of Europe's largest research facilities, neutron research facility ESS, which Sweden runs as host country among fifteen nations.

I enter here, through a nondescript door at a side entrance. My unexpected guru from the lecture I attended earlier, Professor Inger Björck, is on sick leave but has directed me to a trusted coworker.

Juscelino Tovar is a biologist from Venezuela. He came to Sweden because it wasn't possible to get a doctorate in nutritional science in his homeland. After completing his dissertation in Sweden, he returned home but then was invited to come to Sweden for good and decided to take this once-in-a-lifetime opportunity.

"Things had become politically unsettled in my country. That was also a factor," he says.

We take so much for granted, I think. Like researchers being able to do their work in peace, without governments confiscating materials or censoring opinions, or people throwing firebombs outside.

Juscelino Tovar is a thoughtful man with a glint of humor in his brown eyes. He was drawn to Inger Björck and her team because of their deeply innovative approach to nutritional research. Unlike the majority of researchers in this field, who tend to study how human beings function when they've already become broken and sick, this little team is thinking about something quite different, which is similar to what the Ayurveda doctors on the sleepy patio in Kerala were thinking about. That's why I'm here.

"We want to see how you can keep healthy people as healthy as possible, for as long as possible," says Juscelino Tovar.

After many years of research, in which Inger Björck studied how poor health can be prevented by various means, the group decided to put all the pieces together, resulting in the revolutionary study that showed that anti-inflammatory foods, combined in the right way, create good health on many different fronts.

The study had forty-four participants. There were nurses, teachers, researchers from the University of Agricultural Sciences in Alnarp, a couple of pensioners, and some participants who simply were bored and wanted to have something to do.

"But did they have anything at all in common?" I ask.

"Yes, they all had a close relative with some kind of illness," Joscelino Tovar says. "So they had an awareness, the desire to try something new."

I feel that I have to lay all my cards on the table, so I tell him that I've been working on "this" for a while.

He looks at me thoughtfully.

"Do you mean that you actually live like this?"

"I don't think I've been eating exactly like your participants—we'll see. But according to pretty much the same general principles. Yes, I live like this," I say.

He looks at me as if I'm an animal at the zoo. Very kindly. But still—a study object.

I can see that he's thinking, "Does she really do this?"

And how is the new lifestyle going, anyway?

It's moving forward. I'm beginning to understand that this lifestyle is not an end destination but a constant journey where every new mile on the road deepens my knowledge and forces me to find new motivation.

I've now learned that on the days when I eat only anti-inflammatory foods I feel incredibly good: lighter in my body, stronger, with a clearer mind. I feel happy and more mentally flexible. My back stays completely pain free.

At the same time, I wrestle with myself sometimes. In some ways it's easy, since I'm not on a reducing diet and I'm not hungry, but full and sated. It's good and enjoyable, my new way of eating. But so much in our daily life is filled with gluten, lactose, and sugar. The traditional food of the farming society, in modern industrialized form, exists everywhere in my life.

It's hard to let go of toasted bread. It's hard not to drink wine. It's hard to always think about what I'm eating. It's tough to plan food when I have so many other things to think about. It requires awareness, being one step ahead. I've understood all this. But then comes a new insight.

It's more challenging to eat really well when I'm eating with people I like a lot. I realize how emotional my eating is. Also, that it's strongly connected to feelings of joy, anxiety, or sometimes just boredom. Food can be a key to so many moods. I'm forced to admit that I'm an emotional eater. When I'm with friends, it's easy to have too much of everything. It's that feeling of being part of the human village, doing what everyone else is doing without thinking about it.

So sometimes I have to let go and forget about my new lifestyle and just, well, let it be a regular life.

But Rita coaches me. Instead of thinking about my happy excesses as steps backward and getting angry with myself when my back starts to ache after just a few days of relapsing into my old eating habits, she teaches me to just say thanks for the pleasure, as in "Thanks, little girls' night out, it was nice to see you," and then just move on without worrying about it, back to my food list as if nothing had happened, with a shrug of the shoulders. She encourages me to think more in terms of an 80/20 method: eat really well 80 percent of the week, and then enjoy Friday wine and a good dessert if I feel like it, in the company of good friends.

Making a food plan has become a solid habit, whether in my notebook or in the new apps that I find. That's how I discover that it's incredibly easy to consume fats and carbohydrates but that it requires an effort to eat as much protein as I need, about one and a half to two times my body weight in grams. I struggle to get enough protein.

I've even started planning in advance to make mornings work. In the evening I put out all the pills I'm going to take and the lemons I'm going to squeeze, as well as mix the dry ingredients for my smoothie. I've done this in the past for my children, but not for myself.

My husband looks at me with the same look as Juscelino Tovar.

"You have to be really motivated to keep going on like you do," he says.

And I'm not always motivated. Sometimes I'm tired, lazy, sluggish, or in a bad mood, and I don't have the energy. Then it's harder to think about long-term gains. I get stuck in what feels comfortable at the moment and excuse it to myself by saying that I'm tired of being so damned self-centered. (Note to self: This is just a bad excuse—as if it's selfish to take responsibility for your health and for being pain-free.)

All of this is drifting through my head as I sit in the room with the scientist. I'm thinking that a study about radical lifestyle changes requires a lot from its participants. Such a transformation of life is easier said than done.

And what foods were the study participants allowed to eat?

The Lund scientists made up a food plan. They designed their study in the same way that new medicines are tested, by setting up two groups: a control group and an anti-inflammatory group. Halfway through, they switched the two groups' food with each other. They didn't change the food products but rather the diet as a whole. Of course, they weren't able to make the study "blind," meaning that the participants themselves wouldn't know which group they belonged to and what food they ate. After all, you can't disguise a salmon as a meatball. The control group was given regular everyday food, based on what's called "Nordic nutrition recommendations."

"It sounds wonderful," I say with some irony.

"The National Board of Health and Welfare," says Tovar, with a nod. "No strange recommendations. No large amounts of sugar. Some omega-3 fatty acids."

"And the anti-inflammatory group?"

"Designing that diet was a big job," says Tovar.

The team scanned their own database with a magnifying glass, covering years of data from earlier studies by Björck and her team. They also plowed through the scientific literature about the connections between food and health, all in order to find leads about which foods could best promote an anti-inflammatory health effect. There was a large body of research that had been done on animals, and there were many speculations. But less research had been done with people. And that was what they were looking for: substances that had a documented anti-inflammatory effect on the species Homo sapiens.

Out of this review emerged five basic principles that had previously been shown to be effective anti-inflammatory strategies, each in a separate study. Now these principles would be combined to form a kind of anti-inflammatory health bomb, in which all of them would work together at the same time. (I'm reminded of the Ayurvedic thinking. This idea, signaled by words like "wholeness" and "synergy," is something that an Indian doctor would recognize immediately, with the enthusiastic head shake that means "yes.")

But the team had a number of obstacles to overcome.

"We had five principles. How could we convert all that into a specific food and lifestyle plan? It was a huge headache."

They had to take the big step from thinking in terms of chemical components to thinking about food. Real food, good food, interesting food, food that could be found in a regular food store on a rainy Tuesday afternoon in November and that could be prepared by ordinary people.

As Juscelino Tovar talks, I'm wondering if the good scientists haven't complicated things unnecessarily, since I know several people nowadays who eat exactly this food every day. But I'm too polite to say anything, at least for the moment.

The next challenge for the scientists was how to get people to follow the program exactly. (What science calls compliance.) It wasn't enough to design menus and present general and well-meaning advice. The researchers also had to be prepared for people to do what people do: go off the track, lose motivation, get tired, cheat a little, or just get mixed up by all this disciplined eating. They decided to establish a telephone hotline where the participants could call in and ask questions about all the problems the new food caused for them.

The menu was detailed, as I said, with exact weight and volume measurements. Men were allowed to eat slightly more calories, women slightly less. The goal was not weight reduction; it was better health.

They recruited the forty-four subjects, who got started. The program wasn't always easy to follow.

"Some of the participants would call our dietician on Fridays and say that they were going to a party now, and what were they supposed to do?"

"Were they ready to give up?"

"We took it as proof of their engagement," says Tovar. "The participants *wanted* to follow the program, they *wanted* to learn more."

And I understand only too well this problem of Fridays.

When I first began my new lifestyle, I also had a hard time going out. I always felt lost about what I could and couldn't eat and spent a lot more time thinking about my food than about all the nice people around me. It felt sad and limiting. This was not me.

"But what are you doing?" a friend asked me. "You don't need to diet."

"This isn't dieting, I just feel better this way."

I met with suspicious looks. I decided that the point of the lifestyle wasn't denial or feeling bad or leading a monastic life. The point was long-term transformation.

I also did a quick calculation. A week has seven breakfasts, seven lunches, seven dinners. Altogether twenty-one meals and ten to fifteen snacks. How many of these do I not decide about myself? At most three or four meals. It's only about one-tenth of the week's meals that I don't have full control over. So I decided that if some well-meaning person had gone to the trouble of making good food for me and my family, I would eat everything, enjoy it fully, and say thank you like a polite person should.

But then I wasn't participating in an advanced research study, whose results would soon stun a doctor in Dalby. Because after four weeks the results come in, and the general practitioner in Dalby calls up the researchers. He can hardly believe his own eyes.

"He asked us what exactly we'd been doing," Tovar remembers.

That's how dramatic the results were.

"He actually refused to believe them."

He falls silent, looks out the window and shakes his head.

"It was really quite amazing."

We're not just talking about the bad cholesterol that decreased by 33 percent, the blood lipids that decreased by 14 percent, or the blood pressure that fell by 8 percent. These are absolutely central markers since they can be linked to "the great people killer," the metabolic syndrome, which we know is connected to diabetes 2 and cardiovascular disease and which we suspect is linked to certain cancers and dementia as well. It was also clear that the program lowered inflammation levels. At the same time something had happened to the participants' mental capacity. It had increased!

In other words, the study demonstrated that inflammation markers were linked to the whole system, to the whole person's well-being. Cholesterol, blood fats, blood pressure, diabetes, heart, veins, mental capacity. The whole thing. If a bestselling author had written a book about the Lund scientists' diet, it might have been titled *Healthier and Smarter—In Just Four Weeks.*

This revolutionary yet modest team in Lund had in other words conducted one of the most interesting studies to be carried out anywhere, about how normal, healthy people could become even healthier and thus more resistant to disease.

Now I'm sitting on the edge of my chair, dying to know exactly what kind of food the participants ate. I want to see if it's "my" food. Or will new things turn up?

I ask about the basic foundations of the program. And for the first time I hear my "Rita Diet" described in scientific language, dressed in a white lab coat.

Principle 1 was to increase the amount of foods containing omega-3. Omega-3 is found in all fatty fish, for example salmon, which is a known omega-3 bomb, and also in less fashionable kinds of fish like mackerel, sardines, and herring. But omega-3 is also found in nuts and seeds, like almonds, walnuts, and chia seeds. This doesn't surprise me at all. The health effects of omega-3 have been known for a long time.

"It was obvious for us to put this in," Tovar says. "So many studies had shown the anti-inflammatory effect of omega-3."

Principle 2 was to control sugar. Food should have a low GI value, that is, a low rating on the glycemic scale that measures how quickly carbohydrates in food are broken down into sugar. People should also completely avoid refined sugar. Not because sugar is high in calories, as the old way of thinking had it, but because sugar drives up the insulin levels and creates inflammation.

But the scientists took this principle one step further. They also recommended, when eating meals with a higher GI value, adding substances that could actively lower the GI effect, or in other words, lessen the body's insulin response to the food. In everyday language, that means that an old favorite from French cuisine, vinegar, came in handy when eating rice or potatoes, for example.

"We even made a special salad dressing," Tovar laughs.

"But why?"

"Vinegar lowers the GI response to any meal," he explains.

And we talk for a while about these French gourmets, who eat so well and like to begin their meal with a salad with vinaigrette—and who manage to look slim and fit into their later years, even though they actually eat substantial meals several times a day.

"Could you say that vinegar increases the health value of a meal?"

"Absolutely, because it lowers the GI level," says Tovar.

"So you think the French have understood this for a long time?"

"Maybe, on an intuitive level."

I like intuitive levels, especially when they can be linked to research. There's more.

Principle 3 was to eat more fiber-rich foods, especially the soluble, viscous fibers found in vegetables and beans. It's been known for a long time that these fibers are good for stabilizing blood sugar levels. This is where berries and fruits also come into play, and the lingonberries that Inger Björck talked about earlier, all of which contain large amounts of polyphenols, the large chemical molecules that exist naturally in plants and that have strong anti-inflammatory and antioxidizing effects.

Principle 4 was to consume more probiotics, or good bacteria. The participants were given a tablet with a probiotic supplement every morning; they had chosen the bacteria family *lactobacillus*.

"We had the best studies about lactobacillus. That's where the most well-documented and clear anti-inflammatory benefits were found," Tovar explains.

But good bacteria need food, and this is where the soluble fibers come in again. Food with lots of soluble fiber becomes bacteria food, or prebiotics. Simplified: lactobacilli eat onion, once the onion has reached the intestine.

Okay, let's review.

- Omega-3 in fish, nuts, and seeds
- Sugar control—or low glycemic response, as Tovar calls it
- Soluble fibers in vegetables, berries, legumes
- Probiotics

My brain ticks off point after point. So far I feel right at home.

But then there's new information.

Before the study, the scientists also looked into a new area of research having to do with something called *beta glucans*. I haven't encountered this before, and I lean forward over my notes.

Lingonberries are the red gold of the forest and are now starting to be exported to Great Britain—as anti-inflammatory magic!

Beta glucans, Tovar explains, are substances found just beneath the outer shell of oat seeds. A grain consists of three parts: a little sprout in the middle, endosperm around it, and the outer shell. The endosperm contains carbohydrates and protein so that the germ can grow, and at its edge are these newly discovered beta glucans that turn out to have a number of magical health effects.

Tovar tells me that similar substances are found in barley and rye, and they can be purchased in concentrated form nowadays, as a kind of flake that can be strewn over yogurt.

It's a lot to take in. I write *"beta glucans!!!"* in my notebook, adding that they can be found in health food stores. (Later I find them there, little rabbit pellets that look like what's left over when you've finished planing your cutting board in carpentry class. They are a little coarse, but not at all unpleasant mixed into your morning yogurt.)

But how does all this work?

"What do you think, if you use your intuition?" I ask Tovar. "Do these different parts of the diet each work separately, according to their own principle? Or are their effects more unified?"

"We definitely believe that we're dealing with food that is multifunctional. Take a blueberry. It isn't just the polyphenols in the berries that decrease inflammation; there might be deeper reasons that we don't quite understand. For example, the polyphenols might act as substrata for the bacteria in the gut."

Substrata is scientific jargon for food. So blueberries are bacteria food?

"Yes, and it turns out that the bacteria begin to produce twelve other substances when they've fed on the polyphenols in the blueberries."

For simplicity's sake, let's call these twelve substances bacteria poop. Some scientists think that it might be these twelve different kinds of bacteria poop that give blueberries their anti-inflammatory effect. The polyphenols thus affect the composition of the bacterial flora in the intestine. It's a chain reaction. Blueberries become food for bacteria, which produce different kinds of bacterial poop, which in turn creates an anti-inflammation effect. A cascade of subtle interactions is put into motion, but we still don't know exactly what they are.

"We are on the absolute frontier of research here," Tovar points out. "We really don't know what's happening. Maybe these twelve substances directly affect the genes."

But even if the scientists don't quite understand how the participants in the study were affected, the participants themselves noticed the difference.

Most of them have wanted to continue with their anti-inflammatory life-style. They simply felt better. And as noted, it wasn't just a physical effect, but just as much a mental one, especially in terms of cognitive function—a slightly evasive scientific way to describe intelligence.

People quite simply got smarter from the food.

A bright, energetic woman enters the room.

Anne Nilsson is a lecturer in the Department of Food Technology, with a background as a civil engineer. She was the one who led the studies about the cognitive ability of the participants and measured how that ability was affected by the new food.

Cognition derives from the Latin word for thought—or as the philosopher Descartes declared so solemnly: *"Cogito ergo sum,"* or "I think, therefore I am," before he froze to death in 1650 in the old drafty castle Tre Kronor in Stockholm, where he had come from France in order to teach Queen Kristina. (In contrast to the motto of martial arts champion Bruce Lee, who stated that "he who thinks too much gets nothing done.")

But regardless of whether you're a thinker or a "doer," we all need to plan our week, learn new things, and put them together into a greater understanding, as well as sometimes solve tricky problems. That's when we use cognition. It's used when doing math problems in school, learning

ANTI-INFLAMMATORY FRUITS AND BERRIES

Choose fruits and berries with a high polyphenol content (colorful and rich in taste) and somewhat lower sugar content.

- Apples
- Apricots
- Black currants
- Blueberries
- Cherries
- Dates—high sugar but very anti-inflammatory
- Gooseberries
- Grapefruit
- Lemons
- Limes
- Melons—especially honeydew and cantaloupe
- Oranges
- Papayas
- Pears
- Pineapples
- Plums
- Raspberries
- Red currants
- Rhubarb
- Strawberries
- Tangerines

Bananas, mangoes, and watermelon are very high in sugar. Try to restrict them mainly to before or after a workout.

new languages, typing on the computer, reading a rental lease for a new apartment, or trying to understand a radio program about who is fighting whom in Syria. In other words, important things related to the higher development of human beings. The basic foundation of a civilized society, you might even say. Without cognitive ability, we wouldn't have shopping carts in the grocery store, penicillin, or makeup tips on YouTube. Nor would we have space travel, democracy, bike racks, or pink iPhones. All of this has been thought out, planned, and troubleshot by humans using their cognitive abilities.

My own cognition is working at full steam right now, as I want to investigate more deeply how this super important ability in human beings is affected by anti-inflammatory food.

Anne Nilsson takes me with her to some little booths in the department. This is where the scientists are setting up cognition experiments like the ones that were done in the large study. What did those experiments look like?

"We read short sentences to our study participants," Nilsson tells me.

It might be something like this. Imagine the scene with a study subject in a booth and a scientist's voice speaking.

The scientist reads sentence one: "Mom is nice."

The scientist reads sentence two: "The chauffeur ate the chair."

The subjects have a short time to judge if the sentences they heard worked or if they were pure nonsense. In the above case, the first sentence is true, but the second one is false, since chauffeurs don't eat chairs.

Then the subjects have to repeat the nouns in the sentences quickly, like "Mom" and "the chauffeur." About sixty sentences like that are read quickly.

I don't quite understand how this could measure something as complex as people's ability to think.

"Yes, this is a good test of working memory," Nilsson says.

She explains that working memory is about being able to focus and concentrate. (The scientists are careful to explain that it's not about pure intelligence, but working memory and intelligence have been connected in other tests.) Here the subjects had to keep the first word in the sentence in their minds in order to see if the rest of the sentence was logical. If they didn't remember that it was a chauffeur at the beginning of the second sentence, they might well think that it was a giant tiger, which might indeed have eaten a chair if it felt like it.

The remarkable thing is that the scientists could already see a positive effect on working memory after the participants had been eating the anti-inflammatory foods for four weeks. Let's repeat that: The largest asset of human beings, our thinking apparatus, could be sharpened significantly by anti-inflammatory foods in four weeks.

Does that mean that kids with problems in school could get a new chance, just with the help of food? That we all could become radically smarter, that society as a whole could solve problems more quickly and more effectively if we switched to a more anti-inflammatory diet?

No wonder that after the study most of the participants wanted to continue their lives as more healthy and informed.

But Nilsson also tells me this:

"The effects were reversed when the participants changed back to standard food again."

So this improved thinking ability needs to be not only restored but also maintained?

In the course of the study, the scientists were also puzzled by what they saw in the case of a few participants who were completely unaffected by the diet, both mentally and physically.

"We can't explain it," says Tovar. "It's strange. The effect was so dramatic for the majority of the participants."

What is the next step for the researchers?

"We have to bring this out of the research environment and create a diet that works," Tovar says.

I tell them that I myself have a diet that looks a lot like this one and that actually can work in daily life. And now I have to ask.

"What exactly am I doing to myself?"

"We really don't know," says Tovar. "But I think that if we combine several different anti-inflammatory mechanisms, we reboot the system."

"You mean that the whole system is reset?" I ask. "Do you mean it's like a rejuvenation cure?"

"I think that possibly people return to an original background level," Tovar speculates.

So human beings might have a kind of healthy original level that is worn down by low-grade inflammation. It's kind of like when an iPhone gets flipped out by having too many apps that use up the battery, so it no longer has the energy to be itself and needs to be recharged.

It's clear what question I have to pose next.

"If so many people feel better when they eat anti-inflammatory food, and if there seems to be a logical explanation of why they feel better, then why isn't this spread more widely among doctors?"

We begin discussing Western medical science. We desperately want to believe in the ultimate cure, in the shape of a pill or a substance that's going to cure us while we continue to live as usual.

"It's hard to break that belief," Tovar says.

"But take diabetes," I say. "Isn't it strange that when there's a disease that's linked to food, doctors don't give better advice about following an anti-inflammatory diet?"

"Doctors don't know what kind of power food can have," says Tovar. "They tell patients to take their medication but have very little knowledge about what food contains and how nutrients are absorbed."

"But how can that be?" I ask with surprise. "A medical education takes five years—shouldn't diet have a bigger place there?"

"Doctors are pressed for time," Tovar says. "They're trained to treat, to take concrete action with fast results."

"Is it that they have too little time with patients?"

"Too little time to work with diet, yes."

Working with diet requires taking a longer view, being goal oriented, and having the ability to engage and educate the patient.

"Are the doctors *unable*? Or are they *unwilling*?"

"When you ask people to change their diet, you get a lot of resistance," Tovar says. "A lot of resistance. It's deeply rooted, this habit of white bread and pasta."

"You have to get this knowledge out into the world," I say.

"We're doing our best. We give lectures at the university and the doctors are interested, but we still end up being a side branch, since we're working on prevention."

Here we might be reaching the heart of the problem.

In our health-care system, there are limited resources for working on prevention, that is, to ensure that people's risk of developing diseases decreases.

I continue thinking. The diseases that I now understand have links to inflammation, like cancer, cardiovascular disease, and psychiatric diseases, make up the bulk of the cost of health care. At the same time, only pocket change is being used on lifestyle, like preventive diet advice.

In other words, when it's a question of patiently explaining to a fifty-five-year-old pizza lover that he could be eating better food, the health-care system has no resources for that, and the doctors have neither the knowledge nor the time. But ten years later, when the pizza lover has an acute heart attack, then there are suddenly resources to spend and space to use. Ten years earlier, the pizza lover could have gotten help with his pizza dependence through good advice. Now instead he has an acute fear of death and a lot of pain, not to mention the suffering and anxiety that his family must endure. The whole logic is faulty.

It's as if we are the world's best at putting out fires, but we build with flammable, dangerous materials, are careless with fire, cook over open flames on the floor, and don't have any fire alarm. Is this logical? Hardly.

This is a system that has caused us and our fellow human beings great suffering, and at the same time it is both primitive and incredibly expensive. When someone is really ill, we rush in with full force; the ambulance sirens wail and the surgery staff works under high pressure. This is where the politicians go in and make grand gestures and huge investments. It's here that doctors become heroes, as they look at their blotchy X-rays with a steely eye. It means that the pharmaceutical companies' innovations are aimed at treating disease rather than preventative therapies, since these companies sell their medications to a society that's mainly focused on those who are already ill. Besides, it's not possible—and let us be plain here—to patent a blueberry. It's hard to make money off the anti-inflammatory diet since it's all about food that's close to nature. Right there, the pharmaceutical companies lose interest since they are in the business of being in a business.

When the public health-care industry doesn't want to entertain the thought that salmon and lingonberries are medicine, and the pharmaceutical companies don't bother about salmon and blueberries since they can't be patented and sold, other visionary actors turn up and take the lead.

That's why I have to turn my gaze in a completely different direction.

Beneath our superficial
differences we are all of us
walking communities of bacteria.
The world shimmers,
a pointillist landscape made
of tiny living beings.

—*Lynn Margulis
and Dorion Sagan*

6. GUT FEELING

The names of the places that I'm driving past sound like something from an Agatha Christie novel: Haslemere, Hindhead, and Devil's Punch Bowl. I'm on my way through Great Britain to the heart of Surrey, the landscape southwest of London. Bright yellow broom is blooming along the roads, and in the woods next to the highway I see riders on cute little ponies.

This is the road out past Guildford, to Grayshott Spa, a formerly somewhat dated and shabby English spa that has recently experienced an unexpected uptick because of a revolutionary way of treating ill health—through the gut, as they say at Grayshott.

Intestinal health, in other words.

I'm not sure what level of intimacy you have with your own gut. My connection is vague, and I have little more than basic knowledge about the twelve-yard-long winding pathway that takes care of nutritional absorption through the intestinal wall and then regulates daily excretion. Sometimes it works well, sometimes not so well. I'm sure many of us have never given much thought to the matter.

And yet . . . Here a disparate group of people gathers every Wednesday, hoping to cure everything from obesity to depression and stomach pain, just by focusing on healing their gut. The way into the body, they think, is via the gut.

The cure takes at least a week, and I've been curious about what this is all about ever since a controversial article was published in the newspaper *Financial Times*. I've contacted the spa in advance to ask if I can participate for just a short time. Grayshott would prefer that you stay two or three weeks, but I don't have that much time in my calendar.

Besides, I'm traveling with a friend who is struggling with a severe illness, and I want to be able to eat together with her, which wouldn't always work since the gut people eat certain meals in isolation. It isn't possible to participate "light." Maybe you have to have certain rules if you're going to carry out what *Financial Times* recently described as "a dramatically different cure that heals the gut, decreases inflammation, and elevates mood."

Grayshott is a large old Victorian building made of light-colored sandstone and bricks that swirl around battlements and little towers. An arch reaches over the entry. The British poet Alfred Tennyson once rented a home here for himself and his family for a few years—the same poet who wrote "Ring out, wild bells," which is always read out loud on Swedish TV on New Year's Eve:

> *Ring out the old, ring in the new,*
> *Ring, happy bells, across the snow:*
> *The year is going, let him go;*
> *Ring out the false, ring in the true.*

In some ways, the poem could have been written for the cautious women who sit in their bathrobes reading magazines in the lovely living room near the lobby, where I'm waiting to register. A few of them look like they need all of it—ringing in the new, ringing out the old, ringing in truth, ringing out the grief that saps the mind.

The lawn outside is an intense green, even though it's only the beginning of February. Through the leaded windows, I see the thick winding hedges of the garden. The smell of earth wafts in through an open window.

After checking in and making a quick visit to the sauna (wearing a bikini—in Great Britain you always wear a bikini in the sauna), I discover the "gut group."

They are sitting expectantly near the teapots and are already confiding in each other.

One woman has had a serious illness and is now trying to recover. An overweight man in his sixties, sadly, has just buried a son who passed away much too soon in some terrible accident, the nature of which is never explained but seems to hang in the air. The man looks resigned. Another woman just wants to lose weight.

They all appear to be in midlife or in the later stages of middle age. They seem weighed down by a sense that the body is on its way down and their mood is in a minor key. In spite of this, a kind of black humor of sadness drifts over the tea table like a liberating breeze.

Now they have invested 1,500 pounds in their guts—or more specifically, in a week filled with gut-healing food, stomach massage, and lectures. What are they going to do? How is this going to work? I ask around carefully but get varying answers.

"There will be lots of seeds, I think," says the lady who's been very ill.

"We'll lose weight," says the man.

The nutritionists at Grayshott believe that our modern guts are out of whack because of antibiotics and other medication, unrelenting high stress levels, and the fact that the animals we eat are raised with hormones and antibiotics. As a result, we suffer from inflammation and hormonal imbalance, which in the long run negatively affects the whole system. We sleep badly, gain weight, and our health and energy levels go down.

That's the bad news.

The good news is that the body, they believe, has a high capacity to heal itself. And that this whole-body healing begins in the gut. Can that be true? Is it logical?

It sounds reasonable that if the body can't absorb nutrients properly, it will gradually lose its vitality. Another way to look at the problem is that we begin our lives—and we're talking about the first days after fertilization—as three layers of tissue, three little sprouts that lie on top of each other and will gradually close up, the way you roll up a jelly roll. The innermost layer is the endodermal tissue. From this core, the entire digestive and intestinal system is formed, along with the liver and lungs. In other words, the most central functions that provide the body with nutrients, oxygen, and cleansing.

This is where life begins; this is the center of the jelly roll. If this central function is disturbed, everything else is disturbed too, like a game of dominoes.

That sounds logical enough.

At Grayshott, they believe that "when you restore digestion, inflammation steadily decreases in the body, which improves the cells' function, healing, and contributes to an increased ability in the body to eliminate toxins."

One of the basic foundations of their cure is a well-prepared, vegetable-rich diet that includes plenty of fats, including cream and butter, as well as organic meat, eggs, chicken, and fish. The motto is: no sugar, no wheat, no caffeine, and no alcohol. The cure is known for the fact that participants are asked not to exercise, except for the occasional stroll in the fresh air or the yoga sessions that are held at the spa and can be considered "relaxation."

Sleep is considered more important, as are the many treatments you receive, all centered on the stomach and massage.

The program has now been running for four years, and Grayshott has reported phenomenal breakthroughs: joint problems have disappeared, cholesterol values have fallen drastically, and participants have been able to decrease the doses of their medicines for diabetes and high blood pressure. They report that the participants lose weight, get better complexions, better hair, more stable blood sugar, a normalized appetite, and increased energy.

I recognize many of these effects from my own journey. But several things seem to be going on here that I haven't encountered before.

I need to understand more.

What are they doing, and how?

Eat more salmon—it's a real omega-3 bomb.

It's twelve o'clock, and the gut group is gathering for their daily lecture. The theme of the day is carbohydrates. A fit British man in a polo shirt welcomes us in an auditorium that fills up, surprisingly enough, to the very last seat. The last arrival has to stand.

Standing room only at a carbohydrate lecture? Unexpected.

People take out their notebooks. The lady next to me writes at the top: *Carbs—bad?* And I suppose that's exactly the question we all have in mind. How bad are carbohydrates, really?

The light goes out. The man begins plowing through a PowerPoint presentation at top speed.

Flash.

An excerpt from the respected British medical journal *The Lancet* that says that the global health burden of poor eating is significantly heavier than from smoking, alcohol, and lack of exercise.

Flash.

Picture of Hippocrates, who tells people to regard food as medicine and medicine as food.

Flash.

A large pile of chips, cookies, sweetened cereal, sodas. Bad carbohydrates.

Flash.

A table with vegetables and fruits in all the colors of the rainbow. Good carbohydrates.

I'm beginning to wonder if I should leave and give my seat to the standing lady. This is feeling like old news. But then I'm surprised.

Flash.

A picture of a lightbulb. Now the man begins to talk about epigenetics, a completely new branch of genetics that I suspect the gut group participants have never heard of. This is brave.

The man explains that it's like turning a light on and off. He demonstrates by using the light switch on the wall.

"Food turns on the good genes," he says, switching on the light.

"Bad food turns on the wrong genes," he and the light say. "Because food instructs the genes."

Epigenetics is a subject that fascinates me right now—but more about that later.

Then there are several flashes with carbohydrates and fats; flash to proteins, hormones, insulin, glucagon. The lady who has been very ill is half-asleep, but the rest of the group looks enlivened. Several of them lean forward and ask questions. I notice something: the lecturer has an unbelievably flat stomach.

Then he begins to talk about watercress, a vegetable that has a very low carbohydrate density but high anti-inflammatory effect.

And my thoughts go unbidden to my mother's aunt Iris, who died recently at age 102. My British grandmother met my Swedish grandfather in London between the wars. Grandma moved to Sweden, while the rest of her family stayed in Great Britain.

Aunt Iris was a charming British lady who had lived through two world wars and supported two children on a secretary's salary; who had soldiered through life with great intelligence, humor, absolute humility, and exemplary good health. When she turned one hundred, and our family was gathered at her daughter's home in Oxford, I asked Iris about the secret to a long life.

"To be at peace with life. To get along with most people."

And she did that—with one exception: the Germans.

"They tried to kill me with bombs—twice," she said. "During the first world war, and during the second one."

At dinner, Iris told us that she was ill only a couple of times during the years between ninety and one hundred.

"Eat more watercress," the doctor had told her.

And so she did, every day until she reached the age of 102 and died, still absolutely sharp, at a care home in Oxford, filled with light and kindness.

I'm not surprised to find Aunt Iris's watercress at the top of the anti-inflammation list and as a symbol of health here at Grayshott.

Hello, Iris, this is for you, I think.

Then the man butchers the official health and food rules that officials are promoting in both Great Britain and the United States.

In Great Britain, the National Health Service recommends that you base your meals mainly on potatoes, bread, rice, and pasta, and also try to eat less fat. In the United States, the Department of Agriculture has similarly made a food pyramid with bread, rice, pasta, and cereals at the base, with the recommendation to eat six to eleven servings of complex carbohydrates every day.

"The official health advice is making us sick," says the man, gazing intently at his audience.

I jot down in my notebook: *Check out the food pyramids and the food advice given by authorities.* What's true? Whom should we believe?

Then it's Sweden's turn.

A Swedish cinnamon bun, to be exact, which becomes a symbol of how quickly sugar goes into the blood and creates an insulin peak, which in turn leads to fat storage.

"Eventually, the body becomes insensitive to the steadily flowing insulin, and when we reach insulin resistance," says the man, shaking his head, "this leads to diabetes 2, which is increasing dramatically."

Avoid Swedish cinnamon buns.

Then the gut group begins asking questions. They're wondering about potatoes (good along with fat and protein), the sweetener stevia (yes), pasta (no), and balloon stomachs (disappear if you eat anti-inflammatory foods).

Someone should give this man a public education medal. He gives the gut group a half-semester's course in nutrition and fills an auditorium over capacity by offering education to people who probably never paid much attention to the word "protein" before.

I share my newfound knowledge with my friend. She and I have a more urgent discussion about doctors and their view of food as medicine—or lack of it.

She's here at Grayshott because, among other things, her kind doctor is on the front line in every way when it comes to medical treatment but also because the health-care professionals she's met haven't shown the same curiosity about using food in the fight against tough illness challenges.

"Why is that so?" I wonder.

"My doctor doesn't have time. He's so overloaded with work that when I come in to see him he's sitting there making comments on my latest lab results, which he is just then opening on the computer as he sits down. He's under a lot of stress, and he tackles the purely factual side first. Food comes later; it has lower status."

We're talking about deep things. Life and death, inflammation and anguish, anti-inflammation and life joy. My friend is experimenting with a number of weapons from the anti-inflammatory arsenal: omega-3, probiotics, a diet with lots of vegetables, and frequent exercise.

"Finally, I see that only I can make myself well. The doctor is my helper at best, but the only thing he can do to cure me is give me medicines that I myself have to take."

We commiserate about the hard edges of life and laugh at all the crazy and wonderful things that it also has to offer. We especially laugh at ourselves and all the strange things we've experienced together.

We reminisce about a dinner at a restaurant on the Riviera, where we happened to sit next to the American actress Pamela Anderson of *Baywatch* fame, and the French waiters did everything they could to get close to her very curvy figure. Their excuses for being near the table became more and more extreme—you can only serve so much food and wine and mineral water—which finally resulted in them coming in with a tape measure and a level and measuring the angles between the floor and the wall for an hour, in the French bureaucratic way. All just to be able to be near Pamela.

Life is crazy, but it's also serious for my friend. We talk a lot about how illness breeds anxiety and feelings of depression and to what degree you can stem anxiety with the right food.

The soul and food are intimately connected.

This is exactly what the nutritionist Stephanie Moore, who helped create and inspire the gut program, realized. She is a tall, energetic woman who started her career as a therapist for anorexics by thinking about connections among the psyche, therapy, and food.

"I saw that there was only so much I could do with therapy," she explains. "So I decided that I needed to learn more about nutrition itself, thinking that I could heal the psyche by working with nutrition."

She earned a master's degree in psychotherapy, became a nutritionist, and now directs Grayshott's gut program. In addition to the balance among protein, good fats, and the right type of carbohydrates, the program has three other secret key factors.

"Here in the West, we see that people's guts become deeply inflamed over the years, as people age. We heal that inflammation when they stay here."

"And how do you do that?"

"We teach our visitors to become gut smart," she says.

"What does a gut smart person know that others don't?"

"That fat is not an enemy, but a friend. We need plenty of proteins that build up our cells. And there's also the value of the daily probiotic."

This is a fast-growing area of research. Moore leans forward, excitement in her voice.

Probiotics, as we know, are synonymous with good bacteria, or bacteria that interact with the microflora in the gut, producing healthy effects. Along with viruses and parasites, these foreign particles make up what is called the microbiome, which exists not only in the gut but also in the mouth, on the skin, in the vagina, and in the lungs, for example.

A number of studies have examined how many foreign cells a human body contains, compared to its own cells. Early studies from the 1970s concluded that the body contained ten times more foreign cells than body cells. The Human Microbiome Project found a few years ago that it's closer to three times more. The latest review, carried out by the respected journal *Nature* as recently as 2016, points to about thirty billion body cells and thirty-nine billion foreign cells in a human being weighing 150 pounds.

Whatever the exact proportions are, we have a huge number of foreign particles inside us. The most common kinds are lactobacillus and other bacteria in the bifida family. But people don't just need good bacteria. It's good to have a certain number of hostile bacteria as well, in order to keep the immune defense in shape.

All of these bacteria are toiling away in our bodies. One professor compared them to a kind of anthill, where they work and work, around the clock, to keep the human being functioning—digesting food, producing vitamins, and getting rid of toxins.

But we modern humans are bacteria-poor.

"Our forefathers, and I'm talking now about hunter-gatherers, had many more different types of bacteria, upward of two thousand different species," says Moore.

"How many do we have?"

"About five hundred."

"But why did that happen?"

"I think there are many reasons. Bad, processed food, too little fat, too many carbohydrates, and all the antibiotics, of course."

PRE- AND PROBIOTICS

- Prebiotics are fibers that serve as fodder for the good bacteria in the large intestine. Garlic, asparagus, nuts, and bananas top the list.
- Probiotics are food and supplements that contain living, good bacteria and yeast fungi. Yogurt is a natural source of probiotics, as are vegetables fermented by lactic acid, but there are also many different kinds available in pill form.
- It's especially effective to combine prebiotics and probiotics in the same meal. Yogurt with nuts and seeds is a perfect combination.

I now observe that Moore also has an unusually flat stomach. Are the bulging stomachs that we develop over the years caused by poor gut health? Is the bulging stomach in fact inflamed?

The latest research shows that one of the main functions of the bacteria might be combating inflammation. (Why am I not surprised?)

Studies in a number of scientific journals single out the extremely active little *Bifidobacterium infantis 35625* for its so-called immune modulating effect.

"What can we do for our impoverished bacteria flora?"

"We can add to it —probiotics and lactic acid. That's what we give our participants. This yields fantastic results," Moore replies.

Since—appropriately enough—I have developed some temporary stomach problems just in time for the interview, she recommends a cure of probiotics and lactic acids. She brings me a carton full of big bottles of a slimy bacteria solution that has to be kept cold, preferably in the refrigerator. I'm supposed to drink some of this every morning.

It looks hard-core, and I really have to think about how I'm going to explain to my family why half of the refrigerator is occupied by six big bottles of something that looks like sperm samples and tastes like rotten buttermilk.

But this might be just what we need. The effect is starting to be scientifically documented.

The American Journal of Clinical Nutrition describes how probiotics can decrease gut inflammation. It's been shown that probiotics have anti-inflammatory effects and that different bacteria naturally have different effects on the immune system. The more types of bacteria there are, the more different types of work the stomach or "anthill" in the stomach can carry out for us.

Now I'm wondering about the mysterious little bowls that are served to the gut group before every meal. What's in them?

"Fermented vegetables. We make our own here at Grayshott. We ferment all kinds of vegetables, like cabbage, beets, carrots, leeks, celery."

"Should we go out and buy cans of sauerkraut?"

"No, avoid those," she says emphatically. "The ones you find in regular grocery stores won't work; they're pasteurized and that means all the good bacteria have died."

Okay. Now we're starting to get closer to the next horizon of knowledge: the difference between good and bad fermentation.

What is this all about?

Fermentation is an ancient technique for keeping food fresh longer, and it's something that human beings have discovered again and again, through the ages, in all the cultures of the world. The main reason is that fermentation consumes the more harmful rotting bacteria in food, which makes it keep longer.

Wine and bread can in some sense be considered our archetypal

TOP HITS: PROBIOTICS

1. **Kefir.** A kind of cultured milk with three times more probiotics than regular yogurt. The name comes from the Turkish *keif*, which means "good feeling."

2. **Natural yogurt.** Avoid types with low fat content or added sugar. Preferably organic.

3. **Kombucha.** The hipster drink that is becoming more widely known. This fermented tea with origins in Korea tastes acidic and is mildly carbonated. Contains several beneficial enzymes, proteins, polyphenols, and vitamins. Try to choose the kind with the least added sugar.

4. **Miso.** This Japanese soup that's starting to find its way to everyday cooking is made from a paste of fermented soybeans, barley, and brown rice. You can buy miso paste in well-supplied grocery stores or Asian food stores and mix it yourself in warm water.

5. **Vegetables fermented with lactic acid.** Check the health food store, or try making your own! There are lots of good recipes on the web.

6. **Pill or capsule supplements.** Make sure you switch types occasionally so that you'll get many different kinds of bacteria.

foodstuffs—at any rate, the archetypes for agricultural societies. In order to make wine, you add yeast to grape juice, a combination that archaeologists have found traces of in caves, graves, and excavations as early as 4,000 BCE. Since ancient times, we've mixed yeast with lukewarm liquid to make our daily bread, which makes the dough swell up invitingly. Our Viking forefathers mixed honey, yeast, and water until it fermented into alcohol and sent them right into a juicy Viking drunkenness. The same principle is used when making beer, but malt is used instead of honey, or sometimes fruit, potatoes, wheat, or rice.

But we aren't discussing wine and beer right now, rather the process of fermentation by lactic acids, which uses the bacteria—mainly lactic acid bacteria—that exist naturally in vegetables.

Moore describes Grayshott's fermentation method to me:

1. Mince vegetables
2. Add coarse salt
3. Pack tightly in a small jar
4. Close it up
5. Let it stand for up to two weeks.

The lactic acid–fermented bacteria will reproduce faster than other bacteria, lowering the pH value, which will get rid of the harmful bacteria.

Meanwhile, a number of new health benefits develop. The lactic acid bacteria, which have multiplied, help the body to absorb vitamins and minerals and decrease the GI value of the vegetables. In addition, lactobacillus has been shown to—yes!—counteract inflammation. An extremely interesting study conducted by scientists from Utrecht University, among others, shows that a certain kind of lactobacillus can even counteract the harmful effects of smoking, by calming the inflammation that smoking triggers in the airways.

Probiotics are what scientists call immunomodulating, or calming and healing. And human beings have been using them for a long time.

Throughout history, people have come up with all kinds of different fermented drinks and foodstuffs.

In Korea we have kombucha, a drink made of fermented tea that's even being introduced in the West, and kimchi, a spicy and tasty kind of sauerkraut with chili. Europe has its traditional sauerkraut, mainly Eastern Europe, but Germany and France as well, for example in the classic dish choucroute. If you eat Japanese food, you often get small pickled vegetables and pickled ginger with your fish. Japanese miso soup is also fermented. Vegans all over the world have discovered tempeh, or fermented soybean cheese. And in Europe we have kefir, the ancient cultured milk with added lactic acid cultures, and yogurt made of cow's, sheep's, and goat milk. And then sourdough.

"Sourdough is good?"

"Yes, if you're going to eat wheat bread, eat sourdough. The acid breaks down gluten. It's less inflammatory."

I realize that I've found Moore's favorite subject. Are these bacteria the explanation for the unusually flat stomachs that I see everywhere at Grayshott among the people leading the gut group?

Moore leads me back to inflammation.

"I believe that gut bacteria break down cytokines and thus directly affect the level of cytokines in the body."

And what is the next secret?

When Moore says the next word, secret number two, I'm reminded of something I read a long time ago.

In South Africa, a big pile of food remains was found. It might sound trivial—just come over to my kitchen if you want to see food scraps, someone might say. However, these were not just any scraps but the remains of a carcass meal that our ancestors ate about two to three million years ago.

A South African professor, Raymond Dart from the University of the Witwatersrand in Johannesburg, found a small fossil at the same time. It was that of an ancient human, probably three years old, who lay there surrounded by a sea of crushed bones and shells. Professor Dart called the human species *Australopithecus africanus* and concluded that the little child had belonged to a group of cave-people whom he described as

"Are the bulging stomachs that we develop over the years caused by poor gut health? Is the bulging stomach in fact inflamed?"

"hunting, meat-eating, shell-crushing, and bone-breaking." In their earliest history, human beings seem to have spent a considerable amount of time crushing the bones of the animals they had hunted or, even more conveniently, seeking out the cadavers of animals killed by leopards and tigers and crushing their bones.

What were they looking for?

Bone marrow. The marrow contains an uncommonly large concentration of vitamins, along with something that's in short supply on the savanna: fat.

When you look at the early tools that have been found in the Olduvai Gorge in Kenya, for example, you can see that there are hammer-like objects as big as a fist, along with sharp stone knives and wedges that our ancestors used to dig out this delicacy. (They also seem to have eaten quite a bit of brain, but we'll leave that for now.)

The better tools people had, and the more easily they could access the nutrients from the whole animal and from all the nooks and crannies of the animals, the bigger brains they could develop. They could obtain more nutritionally dense food, which could sustain them and support their growth.

Right now we're not in Africa, however, but in Grayshott; and we're going to find out the secrets behind the anti-inflammatory gut diet.

There's one more secret that dates from the early history of humankind but is served up here in modern form.

The name of this secret: bone broth.

Bone broth is made from animal bones, including the bone marrow. First, you roast the bones in the oven to give them more taste. Then you cook the broth for at least eight hours, preferably longer. At that point the bones begin to give off a number of good things, like gelatin, collagen, and proline. Bone marrow also contains a very large number of fat-soluble vitamins. Bone broth has become a big thing in the health world recently, thanks to its high protein content and good fats, vitamins, and minerals.

At the University of Nebraska, people have studied bone broth made from chicken bones, which has been proved to strengthen the immune system. In health food stores, there are expensive solutions that are supposed to strengthen the joints. They often contain chondroitin sulphate and glucosamine, which seems to modulate the immune system. But bone marrow contains these elements naturally.

Then I hear a lady say:

"We're only getting bone broth tonight, and then some tea tomorrow morning. No real food until lunch tomorrow," she says wistfully.

And here's the third secret: fasting.

This is not fasting in the way that was popular at spas in Sweden twenty to twenty-five years ago, however. (A friend of mine told me about a spa where he was supposed to drink herbal teas and diluted apricot juice for two weeks, and one of his few amusements had to do with the personal nozzle he had been given to use with the spa's enema bottle. Everyone in the spa went around with their nozzle in the pocket of their robe, brandishing it now and then when the gang needed to be livened up, saying something like "Here's my nozzle, where's yours?")

This is the new way of fasting, which Rita also introduced me to, called intermittent fasting.

This is the logic behind it: Our ancient ancestors didn't have constant access to food. It was sometimes very hard to find food, and they had to go without for periods of time. Starvation turns on the master regulators, a kind of survival mechanism inside our genes, which in turn emit a cascade of anti-inflammatory principles: repair, fat burning, and so on.

All of us have a period of fasting every twenty-four hours. It's called night. That's when the body concentrates on burning food and then getting rid of waste products via the liver and kidneys. I tell Moore that I do a mini-fast two mornings a week, meaning that I postpone breakfast for a few hours, giving my stomach extra time to digest.

"That's good, but you could do more," she says.

Then she begins telling me about several different methods of fasting. Grayshott has had very good results with the one called 16:8.

That means 16 hours without food and 8 hours with food. It could mean eating an early dinner at around 6:00 PM, and then not eating until 10:00 AM the next day. Another, similar principle is 5:2, which many people consider to be an effective way to lose weight and which also has been shown to decrease inflammation markers. On the other hand, it's hard to do this for the long term, since the fasting days require a fiendish self-discipline and "turn you into a UFO," as a friend described it. I've never been able to fast all day; it doesn't work for my body. But again, doing a little mini-fast is very energizing.

FIND YOUR BEST FAST

There are many different types of mini-fasts that have proved to be effective in decreasing inflammation markers.

- **5:2** consists of five days of regular eating per week, and two days of fasting with a maximum of six hundred calories per day.
- **6:1** consists of six days of regular eating and one day of fasting per week.
- **16:8** means that you fast (only liquid, like water and herbal tea) for sixteen hours, and eat during the other eight. It might mean eating your last meal at 7:00 PM, and then not eating until 11:00 AM the next day. This is done one or two days per week to decrease inflammation and give digestion a break, but it isn't as radical as a whole-day fast.
- **14:10** is the same idea as above, but with a shorter fasting period. My favorite.

Which is exactly what the group at Grayshott does.

It's during this fast that the gut group retreats.

After twenty-four hours, I see the group at lunch again. The food on their plates looks good. There's some kind of meat with marrow, vegetables, and a dark, luscious sauce. Next to their plates are little cups of fermented sauerkraut, as well as a little brew of bitter herbs that is supposed to help digestion.

The man who lost his son now seems to have a unique position in the group, and he's looking more cheerful. The women around him lean forward and listen to everything he says in his well-articulated English. I hear from one of the other guests at the spa that the man supposedly knew someone who knew the late and sainted Princess Diana, which in this target group is heavy ammunition.

But the remarkable thing is that the group as a whole is just looking better all the time. They have smoother faces, clearer eyes, and straighter backs. They seem to have a new energy and vitality.

My friend and I observe that they are actually glowing—the whole bunch of them.

What they're doing at Grayshott seems to be working. Something happens there that apparently decreases inflammation and gives people more vitality and health.

Have they found all the parts that are important for an anti-inflammatory life?

No, I don't think so. Because the course is missing another piece, something that's starting to take a strong hold of my life.

Just do it.

—Nike

7. BREATHE

I haven't been on a workout getaway since I took the train from Stockholm to Gothenburg with my old workout friends in the late 1980s.

My thoughts are racing.

This is either a sign that I'm becoming a geek or an old lady—or it's the best thing I've done in a very long time.

I'm sitting on a plane to Toronto, Canada. I bought the ticket with old frequent-flier points collected over the course of many business trips. It feels like an incredible luxury to travel somewhere just to exercise, and I'm constantly questioning whether it's the best use of money and re-sources. All in all, I'm traveling with some trepidation. I assume that the other participants will be younger, in better shape, and in every way more "appropriate." After all, I've seen the pictures of Rita from her fitness com-petitions. What if I can't manage to keep up?

We land in Toronto in the afternoon, and then I travel on to a city that fittingly enough is called London, farther south, toward the American border and the Great Lakes. I've found a minibus that's going all the way there, and I squeeze in among the Canadians with their warm puffy jack-ets that take up a lot of space in the little seats. It starts to grow dark as we drive out into the countryside, which turns out to be a kind of flat prairie. The windows in the solitary houses glimmer in the dusk.

The chauffeur tells the lady at the front that he's going to get divorced. Two siblings squabble over a ticket. People are talking about Canadian hockey. What am I doing here, anyway? I ask the evening darkness out-side. I get off at a slightly shabby but homey hotel on the outskirts of London. At the entrance is an incredibly fit woman just under thirty, dressed in jeans and a down jacket. Is this what you have to look like around here?

It's hard to find someone to help me at the little table that serves as reception desk, but finally I get a room on the basement level, with wall-to-wall carpet on the floor and sturdy fireproof curtains at the window.

Someone knocks on my door, and when I open it I see a young woman with thick, dark hair. I understand she's one of the participants I'm going to prepare a presentation with.

"Hi, are you the one . . . ?"

"Yes, we're supposed to present together."

We start talking. But something isn't right. When I turn my head away she doesn't hear me. Then I understand. She is partly deaf and needs to read my lips. Beneath the pretty curly hair, I see a hearing aid. She's going to give a lecture about how to overcome your inner resistance.

And then she begins to tell me what she's going to talk about, ordinary words that are made compelling by the disability that she lives with. It's a story about being an outsider because of her hearing impairment and also about the will to keep going and to make herself stronger.

"You have to dare to believe in yourself, even when everything isn't like the norm," she says.

She has tears in her eyes. It's such a big thing for her to step forward with this story. I think she is so brave. Her name doesn't mean anything here, but our names are connected by chance in the strangest way, to form a prayer that's spoken by billions of people around the world, in completely different languages. We become a team, and she makes me braver. She also turns out to be a whiz at electronics when I have problems with my computer, plugs, and other equipment, and I've learned that you have to hold on to people like that.

Now we're going to do some intensive training. We're going to try new things, hard things; it feels good.

You just have to breathe.

I've started to develop a routine for exercise where my goal is to train almost every day, often first thing in the morning. I don't always have time. Sometimes it ends up being at lunchtime, sometimes in the evening.

"I prefer that you work out early in the day," says my guru.

An early workout sets up a rhythm for the day and kickstarts a cocktail of beneficial chemistry. I can feel it. Rita talks about endorphins. There's that, yes, but other things too . . . blood sugar levels, satiety, general energy, and even mood.

My workout, which over the years had degenerated into a little unfocused run here, some weight training there, and a little yoga now and then, is transformed into a goal-oriented journey.

Rita sends me my schedules by email. There's a new schedule every month, with four or five different types of workout days every week. I begin to learn to plan for those too. It feels professional.

For example, a week might have two days of lower body training, two days of upper body training, a circuit day, and a day of aerobic and core training. There's less aerobic training than I'm used to. Still, my body shape changes. My shoulders get broader, my waist narrower. My whole body firms up, and I can feel that the proportion of fat to muscle is changing.

I'm careful to increase the weights very gradually, always worried that my back will start giving me trouble. But knock on wood—one day I wake up and feel that something is missing: my back pain! It's like a miracle.

The new stronger Maria.

"If you're going to have a personal trainer, don't you want someone who'll stand next to you and pep you up?" wonders a friend who loves her trainer, someone who does just that.

You should never underestimate a live personal trainer. The physical presence of another knowledgeable person who can support and lead your training has great value for many people. But for me, where I am right now, freedom turns out to be wonderful—the ability to train when and where I want to during days that otherwise are filled with musts. Still, I don't train in a random way but with greater knowledge and more of a system.

This more systematic way of working out also changes my gym persona, if there is such an expression. I learn to be proactive; I have to own the knowledge about my session. I can't disconnect my brain and just do what someone tells me to but have to arrive at the gym with a definite idea of what's going to happen, which requires a certain amount of planning ahead. Once I'm there, I skip the social chitchat, nowadays working out with big headphones and my own music turned up high, like a focused athlete (ha!) with my full attention on my exercises.

I also learn the structure of a workout session, with sets and repetitions, and that different types of workouts give different results. All the new exercises create a completely new space with fun things to look at and try out.

The Iron Cross and Skull Crusher exercises taste like strength. Walking Lunge and Chest Flies sound happy. Then there are all the exercises with an Eastern European flavor: Bulgarian Split Squats, Romanian Deadlifts. I can mentally picture the Olympic broadcasts of my childhood, with hairy weight lifters who swear and groan as they lift incredibly heavy weights. I do a kind of squat that's called the Lumberjack Squat. Rita's training certainly isn't anything for wimps.

Then we have the Arnold Press, named for the bodybuilder Arnold Schwarzenegger. When I lift two weights above my head, bring them down to head level, twist my hands, and let the weights collide with a triumphant bang in front of my chest, I smile at the thought that this exercise was invented when I was a little girl, in the 1960s, by an Austrian muscle builder on Muscle Beach in Venice, California. Here is a fifty-plus woman who's had back pain, doing the same thing forty years later. Well, why not?

Then we have my nemesis, the pull-up, where you're supposed to hang from a bar and pull up your own body weight. It just doesn't work.

I email Rita a message that I simply can't do it, explaining that even though I've built broader shoulders and become a little stronger, when all is said and done I'm from an old bookish family of priests with very narrow shoulders and that no one in my family could ever do a pull-up. And surely no one can expect that from me, menopausal as I am; also I have a pear-shaped body with a very low center of gravity. Isn't that kind of body extra hard to lift, and don't I have very low doses of muscle-building testosterone?

Excuses, excuses. . . . Rita seems unruffled by my explanations. She tells me to continue trying, using the thick straps that you hang from the bar and put your feet in, to make lifting easier.

Just keep trying, she writes.

I'm about to learn to let go of all bad excuses, some kind of female guilt feeling combined with poor self-confidence, combined with not wanting to take responsibility. Instead, I decide to try diving right through the wave and meeting the problems head-on. Try as hard as I can, take one step at a time, and celebrate everything I'm learning.

Not once during the training with Rita do we discuss calories. It's liberating, especially considering that at every gym I go to, I see women and men who work out frenetically, hour after hour, running, biking, rowing, and stepping on Stairmasters. I think of some Swedish poet who wrote about "those who walk joylessly . . ." Who was that, anyway? It doesn't matter—in any case, it just looks dead boring. That's not where Rita wants me to be, grimly exercising just to burn calories.

What kind of training is best? I'm wondering.

Rita's answer to me has been "the one that gets done." But her program combines tough muscle-building exercises with small spurts of aerobic workouts. That's how I learn the word HIIT—high-intensive interval training. And Tabata, short twenty-second intervals with ten seconds of rest in between, a method originally developed for the Japanese Olympic team in the 1970s. Four minutes of these intervals with running, biking, or squat jumps give the heart a real boost. She also wants me to do yoga, walk in the woods, enjoy life, and breathe. Active rest, she calls this, which I suspect is the opposite of lying on the couch, binge-watching Danish crime series.

It's much, much more than I'm used to. It's hard to make it all work. And that's exactly why I've come here, to learn more and to be inspired by how other ordinary exercisers make the day work.

We eat breakfast together cautiously, as strangers do, sizing each other up. There seems to be a nice group of women at the camp. They are mostly Canadians and Americans, with different backgrounds and reasons for coming here. There's a future nurse; a little farther away sits a financial analyst; next to her a housewife, an administrator, an elementary school teacher, a woman who works for an aid organization, and a spokesperson at the big hospital near our hotel.

By the time breakfast is over, we've left all these formal work facades behind and are in full swing talking about our hobby, this new lifestyle change.

"Do you also have trouble finding time to exercise in the morning?"

"How do you manage breakfast when you're traveling?"

Shared experiences help us bond.

It's the fall of 2013, and I've never seen a training gym like the one we're taken to, since functional training hasn't quite made it to Europe yet. It looks like a garage, big and empty, with iron equipment and big car tires along the walls.

It's exciting to get to meet Rita in person. She's a blond hurricane who, in spite of her magazine covers and bikini-model status, is without affectation, warm—a real human being. It's especially exciting to watch her perform the exercises, with energy, intelligence, and speed.

We do circuit training, which naturally includes pull-ups. It's just as I feared. I'm the worst one, the only one in the room who can't do a single pull-up, not even when I use the thick elastic bands that are supposed to help. When it's my turn, Rita jumps up on the equipment next to me and helps me, and I do a little better. I realize that you really don't need to follow the script of your old family stories; you can write a new one yourself. You can be the worst at doing something and still be the best one at trying.

We have to push something that looks basically like an iron sled with weights. It's unbelievably heavy. Two by two, we turn over giant tractor tires, which is incredibly fun. We work out, banging around with our ropes, rubber hoses, iron bars, and boxing gloves.

In the evening we prepare "Rita food" together and eat it around a big table. Like one big heart, we embrace each other, sharing the big stories of our lives. It's as if the training we've done together with tires and iron sleds has opened the channels among us, moving us swiftly beyond polite surface interactions to a deeper connection.

When I call home and tell my husband that we've been crying together for hours, about each other's old sorrows and hard traumas, about car accidents, divorces, mental illness, alcoholism, and vulnerability, he wonders what the heck is going on.

"You've only known each other for a few hours," he says with surprise.

Or else we've known each other all our lives. We're somehow all on the same journey, and the training is a central key to this new space. Why does it do us so much good?

Some time later, I'm driving over the Öresund Bridge between Sweden and Denmark. The bridge rises in a soft arch over the blue, misty sound. There the city lies waiting: Copenhagen, with its bohemian sexiness, its old, elegant, urban facades, and its airy, modern architecture. You can breathe here, I think to myself. And who could be a better guide, when it comes to breathing and exercise, than the person I'm going to meet?

My son, who is studying in Copenhagen, has begun to focus on exercise and is becoming quite knowledgeable about it. He tells me about a fascinating person he saw on a morning show on Danish TV. Could she be the one I'm looking for, the one who can describe in more detail how exercise affects inflammation?

Professor Bente Klarlund Pedersen is a leading expert, pioneer, and innovative thinker about the border zone between exercise and immunology. She writes a weekly column in the Danish paper *Politikken*, and she's conducting research that focuses to an unusual extent on the everyday health of ordinary people.

A woman in her forties, wearing a long, stylish jacket and jeans, welcomes me at the Rigshospital, on the top floor of the modern building. She has curious brown eyes that glitter under her straight-cut bangs. We get coffee in the research cafeteria, where a big group of students is sitting over lunch. Professor Klarlund Pedersen talks as we walk, balancing our full coffee cups, explaining that she lives as she teaches. Every morning, she runs around the three lovely lakes that are situated right in the center of Copenhagen.

"It is *dejligt* [beautiful]!" she says.

I explain to her that I'm a little confused. I have a feeling that exercise reduces inflammation, which is partly based on the changes I've observed in myself, partly because I know that exercise counteracts the very problems that are linked to inflammation: cancer, diabetes 2, cardiovascular disease, and obesity. But I'm finding contradictory information when I try to understand the big picture, so I'm looking for more facts about how exercise affects inflammation.

"I can't make it fit together, and I'm quite mystified," I say.

"Okay, so let's take it from the beginning," says Klarlund Pedersen in her perfect Danish-Swedish, and smiles.

The professor became interested in the connection between exercise and immune defense while she was working on her PhD and observed that regular physical activity could affect the immune system. This was also the case, surprisingly enough, for people who were completely paralyzed from the neck down. When the muscles of these people were stimulated electrically, the muscles began to produce certain substances even though the brain could no longer control their movement. The scientists eventually concluded that the muscles seemed to have an immune defense of their own, which was activated when the muscles began to work.

"We're talking about a completely separate system. We called the proteins that were produced myokines," she says.

"So muscles can communicate?"

"Yes, we've begun to understand that. Muscles are not dead but very much in contact with the rest of the body."

This communication takes place through a number of mechanisms, but the main conversation happens via a factor called IL-6, interleukin 6, which is produced in large amounts as soon as you begin to move. My problem, or the thing I find confusing about this, is that I'm coming across this very same factor in other places, where it's considered to drive the inflammation up—in other words, in a harmful direction.

How does this work? How can IL-6 drive inflammation in some cases, but when it's released during exercise, it counteracts inflammation instead?

"You have to look at the whole picture," says the professor.

And she begins to draw.

First, she draws a curve showing how exercise triggers the release of IL-6, which in turn triggers a number of anti-inflammatory substances.

She then draws another curve, where an injury or infection kick-starts IL-6 at rocket speed, along with other inflammation-causing substances. But in this case, the anti-inflammatory substances that you see with muscle work are not there.

It's the same substance, but it has a different effect?

"Why has nature used the same substance?" I wonder.

"Maybe nature is economical and inventive," she says with a laugh.

"So how do you see the relationship between exercise and inflammation?"

"My interpretation—and our research shows this as well—is that exercise is incredibly useful for lowering low-grade inflammation, specifically the low-grade inflammation that makes us sick."

"And how does that work?"

"It works in two ways. First, you have IL-6, which triggers other anti-inflammatory substances, like in the drawing I just made. But then there's another effect. IL-6 also activates visceral fat."

Visceral fat. I smile to myself. A few years ago, we measured the fat composition of selected members of the family. Visceral fat had proved to be incredibly important, according to the personal trainer who made the measurements. That's the fat that's stored in the abdomen, around the organs. A person who has too much of it tends to look like—okay, let's be blunt—a seal, with a cute little stomach. This later became an inside joke in the family, whenever someone seemed to be developing more of a seal shape . . . Let's not mention any names!

But visceral fat has many names, from the medical term "abdominal fat" to what we call a beer belly in a man and an apple shape in a woman. This type of human fat is also an active tissue that "communicates" with its environment, just like muscles do. Visceral fat "talks" inflammation more than other kinds of fat.

"This harmful fat is mobilized as soon as you begin to exercise," Professor Klarlund Pedersen continues. "And when visceral fat decreases, then inflammation also decreases in the long run."

I attempt to summarize.

"Then if I've understood correctly, it works like this: Exercise gets the muscles' IL-6 going, which triggers the anti-inflammatory substances and, in that way, also pushes fat out into the blood?"

"Correct," she says. "The visceral fat is mobilized as energy when you exercise."

"Does this happen every time you exercise?"

"Every time. That's why you should move every day—in order to get this anti-inflammatory effect."

I begin to understand that I've become an IL-6 addict who gets my fix every day.

"But you can observe other things as well," she continues. "For example, people who don't exercise have more of the "harmful" IL-6, even at rest."

Wait, this is interesting.

"What exactly are you saying?"

"That inactivity causes low-grade inflammation."

Aha. It isn't just that exercise decreases inflammation. Sitting still also directly increases inflammation.

Professor Klarlund Pedersen has contributed to several TV programs as a kind of fitness doctor. In connection with the program *Praxis* on Danish channel TV2, she was assigned the task of pepping up a Danish family from Fraugde, a small village outside the town of Odense, and trying to get them to exercise.

The family preferred taking the car to biking and the elevator to the stairs, and they liked watching TV at night. In short, they were a regular family who worked and went about their lives as best they could. What she measured was their ability to handle sugar in food. It turned out that this ability improved significantly when they began to exercise and that the insulin peaks they used to get also became lower.

Klarlund Pedersen has described the experiment in her book *andheden om sundhed* (The Truth about Health), and she writes as follows: "The main thing that determines how much blood sugar and insulin rise after a meal is the muscles' sensitivity to insulin."

Muscles have receptors for insulin, and the muscle mass is active in breaking down sugar. This is where it starts to get really interesting again, because high blood sugar levels drive inflammation. Low blood sugar values heal inflammation. And exercise creates lower blood sugar levels.

The ability to process blood sugar is thus the third way that exercise decreases inflammation.

In addition, a person who is physically inactive even for only a couple of weeks loses the ability to process sugar. Professor Klarlund Pedersen and her team demonstrated this in an experiment with a group of physically active young men. Normally they logged ten thousand steps a day on the pedometer, but during the trial, they were only allowed to walk one thousand five hundred steps per day for fourteen days. No other changes were made.

In that short time, the young men lost an average of 2.6 pounds of weight, and their ability to process sugar decreased. In addition, they began to have poorer results on a cognitive test that measured logical and quick thinking.

I think about the research in Lund.

Are we on the same trail again: inflammation and cognitive ability?

I also think about the fact that Rita has been absolutely right about this. You have to be on top of exercising ALL the time . . . #exerciseevery damnday #gymrat #keep exercising.

It's all about strengthening your healthy body and also about fighting bad health.

Professor Klarlund Pedersen digs around on her shelves.

"Now I have to show you something."

She produces some pictures of mouse tumors and tells me that they shrank to half their size when the mice got to run on a little exercise wheel.

"Exercise made the tumors shrink," she explains.

Klarlund Pedersen also shows me new, still unpublished research that shows that when diabetes 2 patients start exercising for an hour per day, more than 50 percent of them are able to stop taking medication.

Can diabetics avoid the need for medication if they exercise?

"This information must be a catastrophe for pharmaceutical companies," I observe.

"Of course," she says. "We've sent it to a very well-known scientific journal, but they didn't want to publish it. They think it's too radical."

"Too radical?"

"They claim diabetes 2 patients are unable to exercise for an hour a day. But we've just shown that they can do it, with the right support. So now we've sent the article along to another journal. We'll see if they dare publish it."

"So why do doctors have such a hard time accepting exercise, and even food, as therapy?" I wonder, frustrated.

"It's easier to take a pill," she says. "Easier for the doctor. There are also strong commercial interests in the pharmaceutical industry who put up a kind of barrier against new knowledge."

We need to digest this together. Is it true that we can't trust the scientific establishment to want to spread the most relevant knowledge to people who are sick?

"It's quite simply horrific," says Klarlund Pedersen. "And that's why we built this center, to counteract disease before it manifests."

On one level, this is a tragedy, since research is now producing more and more results about the terrible cost of inactivity. I find several studies that confirm what Klarlund Pederson has been talking about, that inactivity drives inflammation. But I also discover that this process is more pronounced in women. In other words, sitting still affects women more negatively than men.

There are scientists who speculate that the explanation for this phenomenon might be that men have more hidden motion built into their daily life, while for women, sitting still really means not moving.

Another study of diabetics shows the same thing—sitting still causes inflammation. But it also shows that for every hour that the diabetics exercised (above the average level for the population), inflammation decreased by 24 percent. The evidence is irrefutable.

So why are no changes being made in health care?

Perhaps it's because lifestyle changes are complicated and require so much from everyone involved. How can doctors and physical therapists find the energy to incorporate all this new knowledge and then guide each diabetic on the laborious path toward getting active?

It also requires the participants to be motivated and to keep on working, with all their setbacks, learning curves, and shoddy excuses, just as I have to do myself. And that requires coaches and everyday encouragement—but not everyone has a Rita or a Bente Klarlund Pedersen. Doctors simply don't have the time, and health-care professionals don't have enough knowledge. The health-care system isn't built for this lifestyle approach.

During the interview, I think about the things I've learned about lifestyle change. I'm thinking that the exercise journey is just that—a journey, where there isn't any goal in the shape of an absolute final level that you keep up all the time. It's a steadily ongoing process in which there are victories but also clear setbacks.

For example, my back pain brutally returns when I'm working with kettlebells and try to swing the weight quickly upward. I must draw conclusions from what I can and cannot do. This exercise is not, and never will be, the thing for me. I work hard to lift more and heavier weights, and suddenly manage to break through a weight barrier on my earlier so-dreaded dead lifts when I go up to over 130 pounds—but I push myself too hard and get a pain in my hip. I have to keep still for a month, lifting only little mosquito weights.

It's time to start from the beginning again.

MY BEST STRENGTHENING EXERCISES

- Squats with one-armed shoulder press—for the legs, buttocks, waist, and shoulders
- Push-ups on the toes or knees—for the chest, abdomen, and triceps
- Lateral pulldown—strengthens the back and improves posture
- "Mountain climbers" for the trunk, arms, legs, and heart (and more fun than the plank)
- Bulgarian split squats—perfect for the buttocks and the back of the thighs

Although I get sick much less often than before, even in this new life I sometimes come down with a bad cold or a urinary tract infection. I can become exhausted from deadlines with clients and working at night or weighed down by the challenges that occur throughout the life of a family. During one hot summer, my mother becomes gravely ill and dies, and I lose all motivation to exercise. Back to square one and trying to carefully rebuild my strength.

But for every round, my self-knowledge increases. And I learn to feel out what the body needs just that day. Fresh air? Yoga? Heavy weights and slow training? Light weights, whole body? Running? Tabata in the park?

I have a general plan for the week but learn to be flexible and listen to my inner voice.

Another thing that I learn, which is confirmed again by the professor, is to do as Nike says in their iconic ads: *Just do it*. Make sure you get some exercise in every day. Even if you don't feel like you have the time or energy, or when you feel like doing something else, you can just do it. Try ten minutes of something, and you might feel better after a little while. The only exception is a sore throat; I never exercise when I have a sore throat.

The third thing I learn is about children and how habits are contagious.

Ever since the children were little, I've tried, like all parents, to get them to exercise, eat right, and spend less time in front of all the screens that magnetize and seduce them. I've nagged and badgered and worked with this.

Now, when I'm learning so many new things and have so many new tricks, I'm working hard on myself not to preach at home. I halfway succeed, and sometimes I nag too much and get chewed out by people around me who think I'm driving them too hard and that I'm being a control freak.

Okay, sometimes they're right, like when I read about a study that shows that the most effective exercise for reducing inflammation is to mix aerobic exercise with muscle-strengthening exercises. Running isn't enough, you have to train your muscles as well—which I point out to my husband, who at the moment is obsessed with running and doesn't like gyms.

"Gyms are just not my thing," he says irritably when I preach.

"But the muscle growth sends signals that lower inflammation," I continue, with an even shriller voice. "And after only a month with significantly increased motion, you can cut certain inflammation markers by one-half."

"Look, Maria, just because you've discovered this and it makes you feel good, don't push your whole program on me," he says calmly, walking away.

But the old truth holds: People don't do what you say, but what you do. When I inspire the most, it's because I *do* things, joyfully, instead of nagging.

Like when I make cool new food that tastes good, and it just happens to be anti-inflammatory. And when I set out in my gym clothes and come home happier, or run through a session on the mat and suddenly one of them wants to join me. Then the ball gets rolling and they ask me about food supplements and exercises.

I explain what I'm eating before exercise, with protein, fat, and fruits, drinking coffee and taking BCAA supplements to boost my exercise energy. And how you have to supplement with a protein-carbohydrate combination within an hour after you finish exercising, in order to recharge with new nourishment.

"What does a protein-carbohydrate combination mean?" someone asks.

"Think a couple of eggs and a serving of oatmeal with almond milk."

That's how we've started talking at home, and it's something completely new. The smoothie blender whirs. We test new things—different kinds of protein powder, spinach leaves, blueberries, bodybuilder pancakes, gluten-free oatmeal with lots of seeds and dried fruits, and boiled eggs. The kitchen becomes something of a laboratory, as we crowd around with pots and dishes, tasting and discussing.

My children, adults now, find their own way to relate to food and exercise, and one day I realize that I'm also a pupil now, when they tell me about what they've learned and the unexpected (for me) paths they take. Like my son Jakob, who uses his own mental training to find ways to go to the gym even when he doesn't feel like it. Like my daughter Erica, who's now standing in her mini-kitchen making homemade protein bars, or my daughter Bisse, who goes on long walks in the mountains above Hollywood to get into shape for auditions in Los Angeles. And like my son Gustaf in Copenhagen, who not only becomes a training professional but also discovers the fantastic Klarlund Pedersen for me.

During our interview, the professor suddenly tells me that I must write about smoking.

"Smoking is so incredibly inflammation causing. And it wears down the muscle mass."

I tell her a little more about how I'm searching like a detective to uncover an anti-inflammatory lifestyle and that smoking hasn't been part of my tracking. But of course, Professor Klarlund Pedersen is absolutely right. It's just that I'm not a very effective advocate, since I've never smoked or even wanted to. But of course, everything is related. So let me state here and now that the research is unanimous. Smoking drives inflammation way up.

I also ask about yoga.

"I think yoga uses another mechanism."

"Which?"

"Relaxation, de-stressing . . ."

Yoga has come into my life more and more. I've talked to innovative doctors who believe that patients who practice yoga seem to heal somehow and have better skin. I think of the yoga in India and promise myself to take a closer look at the practice.

But it's time for me to wrap up this interview. Klarlund Pedersen has to return to writing guidelines for the Danish Health Authority. I have a lot to think about and now have found even more motivation to keep myself moving every day.

Then a thought occurs to me. I just have to ask something before I go.

"How old are you?"

"Jeg er tres," she tells me in Danish.

It takes a minute for me to figure out what that means, since Danish numbers are tricky for a Swede. She isn't forty-five, like I thought, but fifteen years older, in other words sixty. She just looks shockingly young for her age. And that's precisely what I've begun thinking about: why the anti-inflammatory lifestyle makes people look younger and have a certain type of skin.

Is the anti-inflammatory lifestyle a hidden youth cure, a disguised skin cure?

I'll just have to keep investigating.

*"People don't do what you say,
but what you do.
When I inspire the most,
it's because I do things,
joyfully, instead of nagging."*

*Skin health reflects the health
of the whole body.*

—Dr. Murad

8. GLOW

It's a crisp spring morning in London. The air is clear and saturated with moisture. A pale pink rosebush is growing along the brick wall next to the urbane boutique hotel in Marylebone where I've made an appointment. I've heard that Dr. Anna Marie Olsen is different from other dermatologists and that she has a unique approach to combating skin diseases and aging.

With a Danish father, who became a decorated British war pilot, and a mother with Greek and Italian parents, she grew up at the intersection of sometimes conflicting cultures.

"Still, it's a very good mixture," she remarks.

It is often said that people who are raised in between different cultures more readily become innovative thinkers, and maybe that's why Dr. Olsen is a different kind of dermatologist. At her clinic on Wimpole Street, where she works in a team with the known French dermatologist and cosmetic-doctor-to-the-stars Dr. Jean-Louis Sebagh, she sees more than five thousand patients every year. And she's begun to see completely new types of results when treating skin diseases and problem skin by using— guess what?—anti-inflammatory strategies.

It seems that food and skin are intimately connected.

That's just what I've begun to notice, and that's why I'm looking up Dr. Olsen.

My skin has a new quality; it feels thicker, with more luster, which I think is linked to my new inner strength. I don't know how to describe it, except with the French expression to "feel good in your own skin." Of course, I don't mean thick elephant skin but a smoother, firmer skin with more luster.

I can see the same thing in other women who are training with Rita and who are my Facebook friends: they all have a glow that seems to come from within. I've started to suspect that skin is very important; it's much more than surface beauty. The skin, in medical terms, is one of the body's purifying organs, along with the liver and kidneys—but it's even more than that.

Skin is self-esteem, a reflection of what's inside, and the tissue that holds our beings together. It's our barrier to the outside world. Maybe it's even related to our integrity—because skin is also relationship, in a tactile and sensual way.

Because I want to understand how all these new things I'm experiencing are linked to inflammation, I've been searching and asking around London to find a truly groundbreaking dermatologist.

That's why I'm sitting here with Dr. Anna Marie Olsen.

"I began thinking about this early on," she says. "Even as a doctoral student, when I was looking at the connections between tattoos and suicide."

"Um . . . tattoos and suicide?"

Dr. Olsen leans forward and takes me through a number of theories that I've never encountered before. One of them is *Moi-peau*, or "My skin," which was developed in the 1970s by the French psychoanalyst Didier Anzieu. Using the language of sociology, and situated within a larger theoretical framework, the theory described how the skin's well-being can be linked to a person's psyche, which summarizes, in much more intelligent terms, some of the thoughts that I've had, on a purely intuitive level.

Interesting.

Dr. Olsen's dissertation was about why people who've gotten tattoos more often commit suicide.

"By what biochemical means could this be transmitted in the body? That was the question I posed."

Back then, this was considered nonsense by many doctors who were more traditional. But today, it has been validated, so to speak, by the research about inflammation.

"The skin and the psyche are both affected strongly by low-grade inflammation," Dr. Olsen says.

She left the tattoo issue and became a dermatologist, but her thoughts and suspicions around inflammation continued to develop. She began collecting information. Clues came from colleagues at scientific conferences who happened to talk behind the scenes about how low-grade inflammation could affect the quality of the skin in general, and skin diseases in particular.

Dr. Olsen looked for additional clues in the medical database PubMed and added bits to her puzzle. She believed she had found proof that gluten and lactose drove inflammation. One of her biggest challenges was the disease rosacea, which afflicts 5 to 10 percent of the adult population. Rosacea sufferers develop large red blotches, like a kind of acne, with blood vessels close to the surface. The face burns and prickles. It can get so bad that the patient retreats from social life.

"It causes a great deal of suffering. And there isn't really any approved treatment. I was groping in the dark, looking for answers. That's why I had to try to find new paths, like food."

She did a study on a group of one hundred patients, where sixty-five of them followed her dietary advice. There was a significant change among the patients.

"It was easy for me, as a doctor, to interpret the results. The patients were no longer bright red. I could see it right away in their faces."

The study has not been published, but Dr. Olsen says that she sees parallel results among other patients—including those who don't have any actual skin problems but simply want to look younger.

"Women come to me and want help with their wrinkles," she tells me. "But what you're really talking about is the aging process and the loss of collagen that takes place in the skin."

Collagen is a protein that supports the body by means of an ingenious construction, a kind of triple helix. Imagine a flexible and strong spiral staircase with three banisters that can stretch out and then twist together again. This staircase is inside the skin and makes it possible for you to stretch it out and have the skin regain its former shape when you let go.

Collagen helps to stabilize important supporting structures in the body like sinews, bones, and skin. As we grow older, the collagen naturally decreases—among other things—when the production of estrogen decreases. Decreased stability in the skin leads to thinner skin and more wrinkles. Research also shows that this collagen is affected by the wrong diet, especially foods rich in sugar.

Sugar and other carbohydrates with a high glycemic index value (white bread, alcohol, pasta, pastries, desserts, and sodas) are rapidly broken down into glucose, which binds to the collagen protein and hangs on to it, forming something called AGE, a type of dysfunctional protein. This more poorly functioning protein easily loses its elasticity and flexibility. When this protein, which is supposed to supply elasticity, loses its springiness, the skin becomes more wrinkled, looser, and more sun-sensitive.

"That's why I ask lots of questions about food. What are people eating? That is a central question," says Dr. Olsen.

Dr. Olsen gives her patients a variety of treatments—from regular medical therapies with conventional medication, to injections of wrinkle-smoothing Botox and the filler hyaluronic acid, as well as therapies that send high-frequency ultrasound waves into the skin. These treatments create tiny burns that in turn stimulate the production of collagen and make it renew itself.

She also performs hair implants for people with severe hair loss, a procedure that in itself entails a risk of inflammation.

"After all, I'm inserting a foreign body into their scalp," she says.

But the hair transplant patients who follow her dietary recommendations have a different response.

"I see significantly fewer irritated scalps."

Given these results, I think that it must be easy for her to convince her patients to change their diets.

"No, it's hard. People are skeptical about food playing any medical role. You've always eaten this way, so why change?"

We talk about how the conventional medical establishment still hasn't realized the importance of diet and that this subject often can feel foreign in the relationship between doctor and patient.

"How do you deal with that?"

"I have to work with the trust they have in me as a doctor, and I ask them to try it out. And when they see results in their skin, in their general health, and their energy levels, then it sticks."

"In other words, the results are your best friends?"

"Absolutely, yes."

"Who are the hardest people to convince?"

"Maybe the psoriasis patients—they're just so tired of all the experiments. But then they see that in only two days they can get rid of all the itching . . ."

With twenty-five patients stepping into her treatment rooms every day, Dr. Olsen has developed a practiced eye.

"When people come into my room, I can see if they're inflamed."

I feel like I've started to see the same thing, although I haven't had the right words for it.

"What do you see in those patients?" I wonder.

"Big, open pores, uneven skin tone, a blotchy complexion, skin that's lost its elasticity."

She thinks for a moment.

"And then there's the issue of wine," she says.

"I can see pretty quickly how much people are drinking," she adds, nodding.

Many of her patients are women who practically live on salad—with wine. The Salad and Wine Diet.

"I have women here who drink three tequilas in one night. And they don't understand that no matter how thin you are, when you drink that much, it's bad news for your inflammation level."

This brings us to alcoholic drinks in general, and in my own life in particular. Having reached this point in Rita's program, I've started to understand more about my own eating and drinking. I can now see two things. Neither one is especially flattering, unfortunately.

1. I'm a social eater and drinker, a people-pleaser when it comes to food and drink, and it's painfully hard for me to make independent choices when I'm with other people if I somehow think it might hurt their feelings. Sometimes I eat food and drink wine to make other people happy, without this necessarily working, and even though no one asked me to. I'm just terribly sensitive about disappointing other people, so I tie myself and my long-term goals in knots—without anyone thanking me for it.

2. I'm alcohol-binary. That is, either I drink nothing, or else several glasses of wine slide down my throat; I think that if I were to line them up, the row would be far too long. Once I get started, I have a late natural stopping point. I rush into it, you could say.

I don't think I have any kind of problem with alcohol. But when Rita asks me to try to abstain completely, I find it . . . hard. Hmm.

Weekdays aren't the problem. From Monday to Thursday, at our regular dinners at home, I rarely drink. But there are many exceptions, and that's when the real challenge begins.

My body itches when I'm standing in the kitchen before a dinner, hungry, and everyone else has a glass of wine in their hand while I have only a glass of mineral water with lemon. It feels empty to drink water on Friday night with my husband. It feels like a letdown to be standing at the theater bar in London, dead sober, when everyone else is having a beer or a glass of champagne, happily chatting about the performance. Or at some work-related party with people I half know, when I have only sparkling water in my glass—it makes it a little harder to make contact.

And I must admit that going to a party wearing makeup and a fancy dress but only drinking four glasses of sparkling water turns me into a slightly buttoned-up and prude person who wants to go home at 10:30 and read a mystery; who, as she takes off her party dress, feels a little sad about everything that didn't happen.

I start to realize that I'm alcohol dependent in social situations; that alcohol is a natural part of how I socialize. In many ways, this is a painful insight. I've seen enough of alcoholism at close quarters to not want to be dependent on alcohol in any way. And now I'm starting to realize that alcohol is really just fermented sugar, which is exactly the thing I want to decrease in my new lifestyle. I also realize that alcohol in itself affects my judgment.

When I've had three or four glasses of wine, my mental super ego disappears—the part that knows what's good for me in the long term. Instead, the little greedy ape peeps out, wanting immediate gratification, right here, right now. The practical result is that I'm digging in the freezer, pulling out the sliced bread and making toast with loads of butter and raspberry jam. Or stopping at the hot dog kiosk on the way home to order a gigantic grilled sausage in a bun with pickles and lots of mustard. Or stopping at the kebab place, or the Lebanese food shop. It's not the end of the world. But that's not how I want to take care of my body and soul.

How should I handle this? On the one hand, there's the knowledge that alcohol really is just fermented sugar/nerve poison/judgment destroyer. On the other hand is the realization that alcohol is part of my life, a pleasant part of the culture that I belong to, in spite of everything. And that I like being a party girl—warm, intimate, and happy.

I discuss it with Rita.

The more I toss around problems with her, the more I discover her unique niche—directness, along with empathy and a creativity worthy of Pippi Longstocking. She has smart views about how to start using alcohol strategically.

"How about if you decide to drink only with people who are really important to you, and say no otherwise?" she asks.

I should use the joy and festive feeling that alcohol can create when I'm with people I really want to experience that with and forget about the rest—the little drinks at work-related occasions and so forth. It's an exciting thought.

I come up with two solutions. One is to start pimping up my water with fruits, berries, basil leaves, and mint. It adds a feeling of celebration, an extra little sparkle.

The other idea is to dilute my wine with sparkling mineral water like the French do, to make a kind of weak spritzer. I pour just an inch or two into the glass, and the rest is mineral water. It gives a cool bubbliness, and I can drink three glasses that way without consuming more than one glass of wine.

Let me say that this way of drinking wine meets with some resistance. People near me snort dismissively, and I encounter questioning, sometimes even aggression.

"You lose the whole point of drinking wine!"

"You're ruining good wine!"

"What a strange way to drink."

We can gently note that drinking is linked to strong feelings and that people don't like changes to their routines. But this new method works for me. I call it a "Rita cocktail." And at the same time, I decide that sometimes it's just time to party, and then I'll use completely different parameters. But then it will be with the people I'm closest to.

If I'm going to lose my footing for a while, it has to be worth it.

Then there's resveratrol, the polyphenol found in red wine that has been shown in several animal studies to have an anti-inflammatory effect. This has inspired lots of magazine articles about how red wine can extend your life and is good for counteracting cardiovascular disease.

"How do you interpret that research?" I ask Dr. Olsen.

"My conclusion is that one or two glasses of red wine a week are okay," she answers. "Not half a bottle a day, though. No, definitely not. But many people drink more than that . . ."

Dr. Olsen thinks for a minute.

"Maybe it's because of stress," she says. "People want quick solutions. Alcohol gives you a feeling of relaxation when you're stressed and temporary energy when you're tired."

We discuss how hard it is for doctors to work with patients on these issues.

"Women think that just because they're slim, they don't have any problems with food and drink."

"Why?"

"Many women have such a fixation on body weight. But they forget that what you eat and drink is so important for health, in ways that go beyond body weight."

"What are they missing then?"

"The fact that inflammation makes you look old," she says. "Uneven skin tone, swelling, wrinkles, loose skin, large pores."

She nods thoughtfully.

"My patients don't understand that their changed skin is largely caused by a constant low-grade inflammation."

Other doctors share her views, like American dermatologist Nicholas Perricone, who has claimed, in a number of *New York Times* bestsellers, that inflammation leads to wrinkles and aged skin and that the best way to rejuvenate the skin is with an anti-inflammatory diet. He has published books where he is pictured on the cover with his firm, taut, glowing skin, holding salmon and strawberries, two anti-inflammatory giants. He has also produced a skincare line containing anti-inflammatory agents for topical use, including a skin cream that I've bought a few times, with incredible results.

This anti-inflammation prophet, Dr. Perricone, draws even stronger conclusions, reaching all the way to the end: he believes that inflammation equals aging, period.

"Do you believe that?" I ask Dr. Olsen.

"Well, I would say that 90 percent of aging depends on inflammation," she answers after a moment. "You should never forget that there is a natural aging process too, a purely genetic aging in the cells. But the skin is affected by both genetics and environment. The right diet slows aging. That's completely obvious."

And now we're approaching the heart of the matter. Anti-inflammation and longevity, inflammation and aging. To understand the whole picture, I have to take a broader approach.

There's an interesting phenomenon I've heard of—but where was that place again?

I click open Google Maps.

"From Monday to Thursday,
at our regular dinners
at home, I rarely drink.
But there are many exceptions,
and that's when the real
challenge begins."

Our bodies are built up from the food we eat.

—Ellen White, founder of the
Seventh-day Adventist movement

9. BLUE ZONE

The mountains near San Bernardino are jagged, with snow-covered peaks that stretch up to the kind of blue sky that you can see only in California.

I'm driving along the ten-lane highway out of Los Angeles. My daughter has been living here for the last two years, fighting like a brave little lion cub to realize her actress dreams in the world of agents and acting studios. While visiting her, I've decided to use this opportunity to take the next step, and so I've carved out a day for this excursion.

Gigantic, clanging trucks rumble like ocean liners, and peppy little Peugeots dart between the lanes. The highway is lined with high power lines, and the cranes in the oil fields are outlined dramatically against the sky. Soon we'll turn off the highway. There's Exit 74.

We turn off toward Loma Linda, a unique place where we will find more pieces of our puzzle.

Short background: In 1905, an older woman and two younger men arrived here after traveling over the Rocky Mountains by horse and wagon. One of the younger men took the woman by the hand and led her down the rickety little steps of the wagon, to the dusty ground.

The woman looked around, took in the palms, the green landscape and mountains, and felt the warm breezes.

"I've been here before," she said.

"No, you haven't," said one of the men.

"Then I've been here in my dreams. And this is where we're staying."

American history is full of these kinds of stories, combining the mythology of the prairie with a sense of inevitability. Like the histories of most other countries, though, it is not so full of genuinely powerful, visionary female leaders.

Ellen White was one such leader. She founded the Seventh-day Adventist Church, which today is a worldwide free church whose believers are different in a variety of ways. They are vegans, pacifists, and among the most long-lived groups of people in the world. These long-lived Adventists are now the object of the whole world's interest.

They form a remarkable zone of longevity, or what researchers have begun to call a Blue Zone. Blue Zones are the areas on earth where people live the longest. Most are on the outskirts of civilization, on peaceful little islands where goatherds and singing farmers putter around calmly among healing herbs—with one exception.

One of these areas, with one single group of people, lies smack in the middle of a big, noisy, dirty, and stressful metropolis. There are impoverished immigrants and drug addicts speaking in tongues; Oscar ceremonies, endless traffic, enormous billionaire villas in Beverly Hills; incredibly ambitious and talented people whose greatest joy is either shopping, success, and money (which leads to neuroses), or artistic expression and a desire to create things in a more experimental way (which also leads to neuroses). This is where my daughter lives, along with thirteen to eighteen million other people—we don't know exactly, since there are so many undocumented people in the city.

In Los Angeles, there are people who practice yoga to thundering rock music, in order to speed up the tempo—the idea being "why do slow yoga when you can do fast yoga?" There are supposedly more therapists than teachers, to take care of all the neuroses; and if they can't help, the ever-present lawyers are always happy to step in.

Here, in the midst of this creeping, smoking stress, lies an oasis. It's called Loma Linda.

Since most modern people don't live on isolated peninsulas or in remote mountain regions, this place might hold additional clues about the ingredients of a powerful anti-inflammatory lifestyle. Inflammation has proved to be deeply interconnected with aging. Even today, scientists can't say which is the chicken and which is the egg—does aging cause inflammation, or is it inflammation that causes aging? Or is it a symbiotic dance in which one feeds the other?

In any case, if people live significantly longer than average here, I'm thinking that they must have found a key to leading an anti-inflammatory lifestyle through their everyday habits. I am looking for these keys.

What do the Seventh-day Adventists actually do in their daily lives? And of all the things they do that are different from what you and I do, which ones are actually significant?

I'm looking for an answer that's buried in a history. And that history began right here, in Loma Linda, when Ellen White arrived in her rickety wagon to found a colony in the warm winds just outside Los Angeles, below the jagged mountains.

That's why I'm standing here in front of a bronze statue of Ellen White and her two sons that depicts the very moment they stepped out of the wagon. The sons are holding their mother by the hand. A bronze plaque next to it describes the divine dream that Ellen White carried with her and how an invisible hand guided her to this place.

Near the statue is the log cabin where White lived when she first came to Loma Linda, while she was founding the colony.

And next to the little cabin is one of the world's foremost universities with a focus on medicine: Loma Linda University, which lies nestled in this suburb. Long, palm-lined boulevards lead through the community, with cozy and simple wooden houses on either side. Loma Linda's school announces that they're having "Brain Awareness Week."

It sounds promising.

The university building is light and open, in a kind of Mexican hacienda style. I'm here to meet with Dr. Gary Fraser, who at the moment is leading the largest study in the world on diet, lifestyle, and health.

"G'day," he greets me.

He's a mild-mannered man in the prime of life. The accent can't be missed.

"Are you a Kiwi?" I ask.

"You can never hide that," he answers.

Of course, he's not the kind of kiwi that's a fuzzy fruit with green pulp and black seeds but rather a Kiwi as in the nickname Brits use for their friends from New Zealand.

Most important, Dr. Fraser is one of the leaders of what's called Adventist Health Studies, the world's biggest study of the lifestyles of the Adventists. Ninety-six thousand church members have been recruited for the study, mainly from the United States and Canada.

"We have a kind of giant natural health experiment that's going on all the time with our members," Dr. Fraser tells me.

The study is led by the scientists at the university in Loma Linda, and they're studying the connections between diet, lifestyle, and the great public health diseases: cancer, cardiovascular disease, and diabetes type 2, using all the techniques of modern science.

The most remarkable thing about these Adventists is that they live so long. For men, it's seven to ten years longer than average for North Americans, and for women, four to five. This high average life span can be seen among Adventists in countries all over the world. They live longer, and develop age-related problems and illnesses later in life, than the populations around them.

"My Adventists die of the same diseases that other Westerners do," Dr. Fraser says. "And the diseases they die of are age-related. It's just that they become ill seven years later than others, on average."

"What does that mean?"

"My intuition tells me that the Adventists have found a way to postpone aging. Something we're doing 'creates youth' in our cells."

"'We'? Are you an Adventist yourself?"

Yes, Dr. Fraser is not just a traditional researcher who observes his subjects like animals in a laboratory. He is also a participant, as someone who is active in the church and as a vegetarian. All of this has played a role in his decision to move to the United States.

"In New Zealand, I lived an hour away from the church and twenty-five other members in my congregation. Here the church is next door and we are six thousand people. That drew me here."

Every morning, before Dr. Fraser goes to the university to study the Adventists' health, he eats his typical Adventist breakfast of whole grain muesli, nuts, berries, and soy milk—a breakfast that the Adventists enjoyed long before the fitness movement, with its kettlebells and raw food, had even been conceived of.

Dr. Fraser is one with his research. He lives and studies the same lifestyle in this airy office, with a view of a lovely pine tree and the blue mountains in the far distance. What is it that drives him, and all the other Adventists, to hold on to their vegetarian lifestyle? How is his belief linked to his diet?

In the 1830s, a strong revivalist movement swept like a wave through the United States. People were waiting for Jesus, absolutely sure that he would show up again—in real life. They had even set a date: on April 5, 1843, he would make his comeback. Ellen White and her parents were part of this hopeful group that lived with a growing feeling of ecstasy before the date. People gathered by the thousands, reformed their ways, prayed, and purified themselves before this return. When the day arrived and Jesus didn't come, there was naturally a certain feeling of disappointment. What to do now? Start looking for gold, open a whiskey saloon, or pray even more?

But Ellen White was not only deeply religious, she was also a doer, and a very constructive doer. She solved the problem of the missing return of Jesus Christ in 1843 by simply pushing the savior's return forward to a vague future time and began building up a new religious movement around this less specific hope. While they waited for Jesus, those who had already died would no longer be completely dead but would remain in an unconscious limbo between life and death.

But that wasn't all.

Ellen White had an idea that her movement, in a more specific way than other Protestant movements, would return to the original rules for living that could found in the Bible. So she began searching the more forgotten chapters and discovered that the day of rest should in fact be celebrated on Saturdays, and not on Sundays as people had done earlier.

Then her husband fell ill but miraculously recovered by means of healthy food and rest. That gave Ellen White another idea.

Her followers would live in a healthier way than everyone else. They would treat the body as a temple, the last gift that God had given them, and she found support for her ideas in the Bible. There Ellen White found God's exhortations in both the third and the fifth books of Moses, Leviticus and Deuteronomy, where God tells his people that certain foods are pure and others impure. In these 2,800-year-old texts, she also found support for how much water an Adventist should drink: exactly six to eight glasses per day, and no more!

She advised the Adventists to fill themselves with the purest, healthiest food that nature could offer, every day, in order to strengthen their bodies and thereby honor God. With every fiber, they would worship nature, live in nature, and offer their families community, healing, and love.

Thus was born the Seventh-day Adventist Church, today one of the world's largest Christian movements, and the only Christian church I know of that's had a female chief ideologist. It's also the only one I know of that sees food, exercise, and lifestyle—in addition to faith and community—as part of the path to salvation.

In some ways, I muse, it might even resemble Ayurveda, which is linked to Hinduism.

But how exactly do the Adventists live?

"We eat vegetarian or even vegan food, and that means large amounts of different kinds of vegetables, at every lunch and dinner. I love vegetables," says Dr. Fraser.

Yes, he loves vegetables. Adventists also like to eat beans, whole-grain bread, fruit, and plenty of nuts. As early as the 1990s, there were studies that showed that Adventists who ate a handful of nuts at least five times a week lived two or three years longer than those who didn't eat nuts. Since then, research has shown that nuts have a beneficial effect on low-grade inflammation, cholesterol levels, blood pressure, and diabetes, all of which can eventually lead to cardiovascular disease.

They don't smoke and don't drink.

But their unique lifestyle doesn't end with their diet.

According to Ellen White, the sabbath, or day of rest, was sacred. She wanted to celebrate it on Saturday, just as the Jews did according to Leviticus, and this day was to be free of work, stress, duties, and obligations. White urged her followers to go out in nature instead and take in the beauty of creation; to walk in the meadows and the forests, and to look at the trees and enjoy the breeze, the flowers, the birds, and beautiful sunsets. Saturday was also set aside for time in church and for undisturbed family time, a practice that the Seventh-day Adventists still maintain today.

She also urged her followers to exercise, even at an advanced age, and there are still special sports clubs for older citizens. For example, there's a swim club at the bathhouse in Loma Linda where the median age is apparently over eighty.

"We are seen as a little odd," Dr. Fraser says.

"Odd?"

"Yes, because of our lifestyle. It's also odd to be vegetarian in the United States."

He has such beautiful skin, Dr. Fraser. His complexion glows and is completely free of wrinkles, even though one might certainly expect a wrinkle or two in middle age. This is someone who could help demonstrate to inflamed patients at Dr. Olsen's skin clinic in London what a skin without inflammation can look like. In fact, Dr. Fraser's whole self is aglow, and something in him exudes goodwill and intelligence.

I think about what I know so far about the link between a low grade of inflammation, increased mental balance, and cognitive ability. I'm beginning to suspect that Dr. Fraser has a very low inflammation level overall, and I just have to ask.

"What about Adventists and inflammation?"

"We have lower levels of C-reactive protein than the average population," he says. "And the vegans have an extra low level."

C-reactive protein, abbreviated CRP, is a protein that's produced by the liver in response to inflammation, and it can be measured directly in the blood. Thus, the CRP level measures the level of inflammation in a person. (A study in *American Clinical Journal* showed that overweight children generally had higher base levels of CRP than children with a normal body weight. Excess weight in itself can generate inflammation, at least in children.)

Among Adventists, CRP levels are lower than among the average population—in other words, they are generally less inflamed. But are there any differences between different groups of Adventists?

Not all Seventh-day Adventists lead exactly the same lifestyle. There are the vegans, who only eat according to Ellen White's original, God-inspired plan: in other words, nuts, seeds, plants, and fruits. There are the lacto-ovo vegetarians, who combine a plant-based diet with milk products, cheese, and eggs. And then there are—shh!—some Adventists who eat meat from poultry, fish, or four-legged creatures.

In the studies, it turns out that the vegans have a lower level of inflammation (lower CRP markers) than lacto-ovo vegetarians, who in turn have lower markers than the fish- and meat-eating Adventists.

But what about meat-eating Adventists—do they really exist?

"Yes, they exist," Dr. Fraser laughs, as if we were talking about people with an embarrassing STD.

"How do you see them? As fallen?"

"Well, maybe a little," he says. "They don't eat that much meat in public, what with the church and so on."

So there might be "fallen" Adventists here in Loma Linda who sneak burgers whenever they can!

"How many are meat eaters?"

"We think that around 5 percent of the people in our study eat meat, poultry, or fish about five or six times a week. But it's also cultural; some groups have roots in countries where you eat a lot of fish."

The Loma Linda study includes a large population with roots in the Caribbean, so it's probably not so much a question of double bacon burgers with extra cheese but rather fish with jambalaya spices.

But even if the so-called meat eating is partly just fish and poultry, Dr. Fraser and his team can see significant differences among the three groups of Adventists—vegans, lacto-ovo vegetarians, and meat eaters. That's apparent in the levels of harmful cholesterol, blood pressure, diabetes type 2, and, as mentioned, inflammation markers.

"We always see that the vegans have the lowest measurements, then come the lacto-ovo-vegetarians. Highest on the scale, with the most risk factors, are the meat eaters."

Take the risk of cancer as an example. If you look at the average rate for Adventists as a whole, the risk of cancer is 33 percent lower than for the general population. Even here, the risk is the lowest among vegans, followed by the lacto-ovo vegetarians, and finally the meat eaters. (Brain tumors, as well as uterine and prostate cancer, are excepted, but when you look at breast cancer or stomach and intestinal cancer, for example, the difference is large.)

Conclusion: The health benefits of the Adventist diet are significant.

At the same time, a program of intensive educational and inspirational work is carried out by the church.

"The church puts great value on how the body is treated, how it was created and should be cared for. That's why we teach our members how to live in a vegetarian way," Dr. Fraser says.

"Does the church teach food preparation?"

"Yes, the church teaches us how to use berries, nuts, soy milk, and whole-grain products to build a stronger body."

It strikes me that our church at home, and all other churches or denominations, ought to come here on a study visit to see how you can take complete hold of the welfare of your congregation. I'm curious about Dr. Fraser's church leaders and how they regard health.

"Are your clergy proud of the fact that you live so long? Do you feel like pioneers?"

"Yes, some of us have started talking about that—Adventists as global health leaders."

It sounds as if the Adventists are the original health movement, and Ellen White, in some unknowable way, came across a number of keys to health; keys that the conventional health movement only found one hundred years later. Or is the health movement in fact a kind of Adventism, minus God?

For real Adventists, that reasoning must seem completely upside down, since the whole point of the exercise is to honor God in the first place.

It seems that different people within the church follow the rules with varying degrees of faithfulness. So which aspect of their lifestyle guidelines lead to longevity? Is it all of them combined, or only some of them?

In order to find out, I have to broaden my horizons and look at the other Blue Zones.

The first Blue Zone to be described was a province in Sardinia, high up in the mountains of the Italian island. There, the journal *Experimental Gerontology* described a study carried out by the scientists Gianni Pes and Michel Poulain. Pes and Poulain had identified an unusual number of people of record-breaking age in the villages of Ogliastra, Barbagia of Ollolai, and Barbagia of Seulo. Between 1996 and 2016, there were twenty people who reached one hundred in the small village of Barbagia of Seulo. These people not only grew old but lived active and good lives, for a longer time than expected. The researchers called the province and these mysterious villages a "Blue Zone"—a long-life zone.

The natural question was: Were these centenarians doing something special? Was there something that these mountain villages offered their inhabitants that didn't exist anywhere else? What were people's usual habits in the region of Nuoro, on Sardinia? Why did they live longer, and stay healthy longer before being afflicted by the problems of old age?

In these villages, there were people who lived the way their forefathers had done for thousands of years. They tended their small gardens, where they grew beans, onions, garlic, artichokes, and fennel—all vegetables that are rich in polyphenol and full of the soluble, viscous fibers that the Lund researchers endorsed. They ate large amounts of capers, which are also anti-inflammatory. They were mountain dwellers, climbing up and down the mountainsides where they herded their flocks of goats.

These goats grazed in chemical-free pastures on the mountain slopes, where they ate large amounts of grass (omega-3) and now and then a small, thin-leaved, yellow herb called the curry plant, or eternelle (in French they call it *immortelle d'italie*, the Italian immortality plant)—which has a strong anti-inflammatory effect. (Its magical effect is described in mythology. When Odysseus, the wandering Greek seafarer, was voyaging around the Mediterranean, he once met a princess who was considered to be as beautiful as a goddess. Every day she would anoint herself with a mysterious oil. Odysseus was worn out from all his years on the ship, but the princess gave him a bottle of the mysterious oil, and presto—he once again became a youthful, handsome man. The princess's oil was made of eternelle.) This was the flower that the Sardinian villagers regularly ingested by way of their goat cheese and goat milk.

And then there was the wine, made from the coarse, dark grapes of the species *cannonau*. Every day the islanders would drink a couple of glasses of this wine, which is known for its anti-inflammatory effect and bitter, rich taste. Not too much, but not too little either.

The researchers asked themselves whether these factors were the secret to longevity. Could you find other zones where people lived in a similar way and definitely state that this was the secret to their long and healthy lives?

Together with the researcher and journalist Dan Buettner, Pes and Poulain began looking for other places on earth with an unusual number of very old inhabitants, with the goal of finding out whether there were any common denominators. They looked for Blue Zones in Asia, Africa, America, Europe, and Oceania.

After a worldwide hunt that involved both epidemiologists and nutritionists, other possible areas began to be identified. The whole undertaking was complicated, since most of the potential Blue Zones were peripheral areas, far away from large cities and well-organized authorities. How could they be sure that people were really as old as they said they were?

Finally, they ended up finding the Blue Zones, one by one.

There turned out to be another zone in the Mediterranean, for example, on the island of Ikaria in Greece. One in three people living there reached age ninety, and dementia, cardiovascular disease, and cancer were significantly less common than on average in Europe.

When I read this, I'm struck by the fact that the people on Ikaria were not only healthier physically—compared with the rest of Greece, only half as many had cardiovascular disease—but they were also mentally and neurologically healthier. They were afflicted less often by depression, and only 20 percent of the islanders over age eighty had any form of dementia, which can be compared to 50 percent of Athenians over eighty.

Their days were spent in everyday walks, physical work, and digging in the garden. They ate fatty Greek yogurt with honey, drank goat's milk, and ate a rich Mediterranean diet with beans, garlic, tomatoes, potatoes, and large amounts of polyphenol-laden olive oil. Every afternoon the islanders drank herbal tea with their friends, using herbs like rosemary, sage, and oregano.

When I see the list of herbs they were using in their tea, it clicks right away. These are the anti-inflammatory herbs, and they have other medicinal values as well.

These islanders were very devout Greek Orthodox believers, and their church calendar encouraged the islanders to observe occasional fasting days during the year. (In animal studies, this type of calorie reduction has been shown to lengthen life and significantly reduce inflammation markers.)

The search continued.

In Asia, they found another Blue Zone. On the Japanese island of Okinawa, otherwise mainly known for the violent battles that took place there during World War II, inhabitants also reached a very old age. They ate large amounts of soy in the form of tofu and miso soup. They used *hara hachi bu*, or the art of stopping eating when the stomach is 80 percent full. These old Japanese would also often plant their own "food pharmacy" garden where they tended and harvested such anti-inflammatory and medicinal plants as turmeric, ginger, and the ancient medicinal plant wormwood, or absinthe, which has followed people for thousands of years. Wormwood has a bitter taste, as anyone who has drunk schnapps spiced with the silvery gray twigs knows.

The Okinawans had developed a taste for bitterness and spiced their rice, for example, with the very bitter fruit known as bitter melon. Many

"I want to dig deeper, even deeper. What is really happening here and in the other Blue Zones? What's happening inside people's bodies? Where exactly does the secret lie? In the gut?"

studies have shown it to reduce blood sugar. In other words, they intuitively lowered the GI level of their rice.

In the Americas, one other Blue Zone was discovered in 2007: the Nicoya Peninsula in Costa Rica. It turned out that the inhabitants there drank large amounts of coffee, ate rice and beans as a main protein source, and talked a lot about their *plan de vida*—life plan, life goal, or meaning of life. They had a strong belief in God and faith that God would arrange their lives for the best.

And then people found the Seventh-day Adventists in Los Angeles.

The circle is closed. That's where we are right now, with Dr. Gary Fraser of the beautiful skin and the interesting research.

What are the common lifestyle factors among the inhabitants of the world's Blue Zones? Researchers are cautious about drawing general conclusions, but there are some patterns nonetheless.

The people get a lot of exercise, but at a moderate pace. On Sardinia, they climb up and down the steep mountain slopes. On Okinawa, the inhabitants work in their little gardens, as they do in Costa Rica. Ikaria is also hilly. And in Loma Linda, people go on nature walks on the weekends. This steady, ongoing everyday exercise is triggered by all the little needs of daily life, except for the Seventh-day Adventists, where the motivation is a religious decree.

The food—and this is common for all the Blue Zones—is rich in vegetables, seeds, nuts, and in some cases, vegetable proteins. The Seventh-day Adventists are the only official vegetarians. In the rest of the zones, meat, poultry, and fish serve as complements to vegetable-rich meals, and people get other proteins from combinations of nuts, seeds, rice, and beans, which together form a perfect protein.

Most people drink no alcohol, or drink with moderation, aside from the Sardinians, who drink quite large quantities of harsh, polyphenol-rich red wine.

The inhabitants in these zones also demonstrate a sense that there is a higher meaning in life—a life goal. This might be called the *plan de vida*, as in Costa Rica, or the Adventist exhortation to take care of one's family, nature, and each other. Or on Okinawa, the wonderful word *ikigai*, which means "a reason to get up yet another day."

Blue Zone inhabitants are social. They mingle warmly and often with family and friends, and care about others. They relax regularly, whether it's drinking herbal tea on the veranda, like on Ikaria, drinking red wine together on Sardinia, or just sitting and visiting with others, like on Okinawa. Or like the Adventists, spending time with the family every Saturday. Time out. Family time. Friend time. Time to just be.

Does this mean that the Blue Zones don't have any problems?

"Isn't there any problem with obesity here?" I ask Dr. Fraser.

"Of course we have obesity—we're talking about Americans, after all," he says. "We have people who drink alcohol as well."

He makes it sound like something very odd.

"Do you drink?"

"I might have a glass once in a while, sporadically, for a special occasion. And I eat fish sometimes."

I want to dig deeper, even deeper. What is really happening here and in the other Blue Zones? What's happening inside people's bodies? Where exactly does the secret lie? In the gut?

It's in order to investigate these questions that Dr. Fraser and his team are conducting a giant fecal testing project to study the microbiome of the Adventists. This is a polite way of saying that the laboratory will be collecting ten thousand little containers of shit from ten thousand

Adventists, which they will examine to see exactly what bacteria it contains.

"Okay, but what is really happening on the cellular level and deep down in the genetic material?" I ask.

"We want to get down to the genetic level; we want to examine how the genetic material changes. We have one hundred and fifty special places in human DNA that we suspect are affected by this lifestyle."

"What exactly is happening there?" I ask.

"We don't know that yet," he says. "But we have our suspicions. Methylation."

"Methylation?"

"I'm afraid I don't have any more time right now—I'm sorry," he says, before I have time to get an answer.

Dr. Fraser has to pick up his wife from LAX. She's arriving from New Zealand, and he implies that Mrs. Fraser doesn't like to wait very long at the airport. He has to get going.

I leave there with the helpful Briana Bird Pastorino, who works at the university's communications department and who helped me book the interview. She isn't an Adventist herself, and I ask her if she thinks they are different from other people.

"They're so humble and friendly," she says. "And they look so healthy."

"And mentally?"

"They're actually more open and flexible than other people," she says.

The cognitive and emotional effects . . . here they are again.

We drive to Loma Linda Market, a supermarket right next to the university; it is run by Adventists, for Adventists. The store is like a giant health food store in a pale fifties color scheme of light blue, pink, and cream. There, along with nuts and produce displays, I see colorful books about

Jesus and Moses, and about Peter, Esther, and Judith. There's a whole section for children, with coloring books of biblical figures.

I continue walking around the shelves and end up among the books about the challenges of daily life: marriage advice, guidebooks for living a loving life, tips for being compassionate, and practical guides for overcoming addiction.

At the juice bar by the entrance, some of Loma Linda's older residents are sitting at a table drinking juice. A pillar is covered with friendly notes with offers from the local church.

> *Are you addicted to food? Come get help and support!*
> *Are you experiencing a marital crisis? Come heal it with us!*

I order a "Zinger" at the juice bar. Carrot, spinach, ginger, and apple whirl around with hemp seeds in an old-fashioned blender.

"Can I have a coffee too?" I ask.

"No, we Adventists don't drink coffee," the guy in the juice bar laughs, embarrassed.

A woman in a long skirt and yellow shirt is standing behind me, waiting for her juice. She tugs at my arm and whispers:

"Actually, we do!"

She was raised in a strict Adventist environment: vegan, no alcohol, no coffee.

"But I drink coffee," she whispers. "Every day."

I leave this Blue Zone with a lot to think about. I'm especially wondering about the common denominator among these people. Exercise, context, food, life tasks, people . . . but those are not the only things that the Blue Zones have in common. There are two more things—two very important things.

My new life is more playful, with anti-inflammatory herbs and spices. Here are some of the most effective ones: ginger, turmeric, thyme, cinnamon, and rosemary.

The first one is the sun. All Blue Zoners like being outdoors. They seek out the sun, sit in it, and enjoy it just like northerners like to do in the springtime, and they like being outside and getting exercise. For many years, we've been living with the advice that sunshine is harmful. But at the right dose it's medicinal. Sunlight helps the body produce vitamin D, which is anti-inflammatory.

The other common factor is greater than the sun.

We're talking about God.

Or rather, spirituality, as it's manifested in the Adventists' church services, in the simple little home altars of the Okinawans, where incense burns, or in a simple Greek church with a blue ceiling and a cross that's silhouetted against the sea. Shared by all these long-lived people is a sense of wonder in the face of the divine, a spirituality that links them to generations before and after them and binds them to their group, their flock.

The journey that began with healing back pain and the feeling of aging has now arrived at God—or Allah, or Shiva . . . you get the idea.

What does God have to do with inflammation?

Now my hunt for knowledge is beginning to take me to places that I really hadn't expected.

What lies behind us and what
lies before us are small matters
compared to what lies within us.

—Henry Stanley Haskins

10. AWE

I've found a new lead and have to continue my journey onward—or rather, back again, to Toronto.

This time, I have a chance to see the dramatic silhouette of the city, with skyscrapers that pierce the sky and the famous CN Tower, which rises like a spire into space, with its futuristic donut near the top. The city is famous for great ice hockey, smoky jazz, and the most amazing oysters to be found along Canada's coastline. It has three million residents.

One of these residents, for the past year, is a new star in the research sky. She even (almost) has the last name "star." I've come across her research, which stands out as completely new and unique. She challenges just about everything the traditional scientific world stands for: the male establishment, stable predictability, and conventions. What else can you say about a young woman who calls herself a "positive emotions detective"?

Behind this New Age title is a well-respected researcher with a PhD in psychology and a deep interest in how psychology is connected to biological markers. She's come to the University of Toronto from Stanford University to do her research.

It's her I've come to see: Dr. Jennifer Stellar. Even the name makes me tingle.

I'm here to talk to her about her revolutionary research on awe. I look in the dictionary and find some synonyms: reverence, deep respect, wonder—the magical and really large things in life. This is what I'm going to interview her about—the meaning of awe in human life, how awe actually decreases inflammation, and what this means for us and how to interpret it.

We meet at the university, where we're able to use a room that works for the TV recording we're going to do.

Dr. Stellar is ethereal, almost otherworldly, with her bright energy; she's pretty and has long blond hair. If I were a Hollywood director planning to make a movie about a psychologist who's pursuing positive feelings in a world of anxiety and aggression, I would give her the role. That's how perfect she is.

I'm eager to get started. But first she leans forward. She has something important to say.

"I'm pretty vulnerable. So I'm very concerned that we make a clear distinction in the interview between the objective findings of our research, on the one hand, and on the other hand, those ideas that are my own speculations and conclusions about what we're doing."

I understand completely. This woman is probably considered a provocation by many of the more traditional scientists in her field. My friend Annette, who's familiar with psychology, called her "a one-eighty-degree-er"— in other words, someone radically different, who projects such light that it brings out the darkness in others.

Dr. Stellar has to be more careful than most people. That's how it can be for creative thinkers who swim against the tide.

Since the interest in psychology first awoke in the twentieth century, triggered by doctors like Sigmund Freud and Carl Jung, psychological research has focused mainly on negative feelings. Why do human beings get angry, frightened, depressed, or suspicious? The positive feelings were taken for granted, as if they just existed.

Dr. Stellar decided at an early stage not to take them for granted.

"Many studies have been done about negative feelings, but I wanted to look at the opposite. What do feelings like joy, pride, awe, and love do to us?"

In a revolutionary study that Dr. Stellar and her team carried out, they measured various human feelings and how they affected the degree of inflammation in the body. These feelings were: love, joy, empathy, pride, amusement, contentment, and awe.

It turned out that of these seven feelings, there were four that decreased inflammation in a statistically convincing way: joy, awe, pride, and contentment.

And of these four, one stood out as an exceptional peak in the curve— awe.

Awe. I'm awed by awe. What is it, exactly?

"If you look in the dictionary, you'll find a number of synonyms," says Dr. Stellar. "The word is hard to capture exactly. But I think we all recognize the feeling of something that's larger than us, something enormous, endless, big. It can be an object, an idea, or a person. You can feel awe when looking at trees or seeing a fantastic nature movie on TV, with a great herd of antelope running across the savanna. You feel it when you look at something that's bigger than yourself, something that blows your mind, something that makes you see the world in a new way."

The idea of linking feelings to immune defense isn't a new one. Another female pioneer, Professor Janice Kiecolt-Glaser at Ohio State University, did notable research in the early 2000s within the new field of psychoneuroimmunology, about how negative feelings affect the body. Among other things, she demonstrated that daily stress, marital problems, and feelings of depression can be linked to an increased risk of inflammation-driven diseases.

So how can you measure feelings and inflammation?

We already know that there are reliable markers, called pro-inflammatory markers, that signal the start of inflammation. There are the so-called cytokines, and one especially, called interleukin 6, or IL-6 for short. This one is the researchers' favorite, for two reasons: it's released quickly, and it can be measured in the saliva.

A summary of the research from Ohio would look something like this: It is now possible to link negative feelings to a variety of diseases. Inflammation has been linked to aging, cardiovascular disease, osteoporosis, arthritis, diabetes type 2, certain forms of cancer, Alzheimer's, and certain dental diseases. And the production of pro-inflammatory cytokines, which drive these illnesses forward, is stimulated by negative feelings and stressful events.

Yet another cool researcher, woman, and pioneer—Professor Margaret Kemeny at the University of California—has shown that the link between emotion and inflammation has as much to do with the feeling experienced as it does with the actual situation. She examined how actors in Los Angeles were affected by their various roles. After performing the emotion "sad" for twenty minutes, the part of the immune defense called NK-cells increased.

(I can't help thinking of the actor Heath Ledger, who played the infernally evil Joker in a Batman movie after preparing himself for months by thinking evil thoughts and living in a completely deteriorated apartment. He later died of an overdose of pills. Was there a link? It's an awful thought.)

"How is all this connected—feelings and inflammation?" I wonder.

"Inflammation affects our mood," Dr. Stellar says. "When your body experiences inflammation, your mood changes. When people are really ill they start to behave differently—an illness behavior. They withdraw, try to take it easy, take time to heal."

I realize that I   and everyone I know—have behaved exactly like that, and I remember my years as a medical assistant, when I would often sit with very ill people. Illness behavior means that we turn inward, withdraw, diminish, and protect ourselves. My dog used to lie in a corner when she didn't feel well, with her pleading brown eyes. So this is inflammation talking to the brain. But which is the chicken and which is the egg?

"Is it my illness that inflames me so that my behavior is changed?" I ask. "Or do my negative thoughts make my illness worse?"

"There's a kind of cycle here," Dr. Stellar says. "Inflammation makes people withdraw. It creates a feeling of isolation, which in turn increases the feeling of depression. But to some extent, this is functional. When you're sick, it's good to pull back in order to heal. You shouldn't be too active."

Nature is wise, in other words. Inflammation seems to steer our behavior toward behavior that's more conducive to healing.

But back to the positive feelings, and above all, awe.

"How often do people feel awe?" I ask.

"Oh, it's a common feeling," Dr. Stellar laughs. "It's not something that happens only when you see the pyramids of Egypt for the first time. In our study, people reported experiencing awe about twice a week."

I can picture the study participants: Awe? Check! One dose of magic, twice a week.

"And what were people doing when they got this feeling?" I wonder.

"They might be listening to a beautiful symphony, or taking a course in astronomy here at the university. They were out in a beautiful landscape, or had a religious experience where they felt the presence of a great and good spirit."

This is what I find so exciting about Dr. Stellar's research. She doesn't record what people think they feel but measures it biologically.

"I don't always trust what people report that they feel," she admits. "That's why I want to find objective ways of measuring it. It's been my specialization as a psychologist—to measure the importance of people's feelings biologically, based on how they are expressed by the autonomous nervous system."

And she finds decreased levels of the cytokine IL-6 in saliva when people experience awe. Which is no small discovery, considering that elevated values of IL-6 can be linked to serious illnesses like depression, cancer, and even schizophrenia. In fact, some of these diseases are treated nowadays by trying to decrease the level of IL-6 in the blood. This is exactly what Dr. Stellar's team has found that people can do on their own—by experiencing awe.

"But are all types of awe equally good, from a purely medical standpoint?"

"We don't know that yet," Dr. Stellar says thoughtfully. "We believe so, but this is the next thing to investigate. There's a common biological component, and it's shared by all human beings."

What could all of this mean?

"Now I'm speculating," continues Dr. Stellar, with a clear reference to our agreement. "I think that one basic principle, where awe is concerned, is that it helps to bind groups together. A human society needs stable groups where people feel interconnected."

"Can you give me an example?" I ask.

"Let's say that you're in a group of people who are watching Usain Bolt run the hundred-meter dash. The awe that all of you feel at that performance binds you together as a group."

It turns out that the room we're sitting in belongs to Dr. Stellar's husband, who is researching political group identity and why political groups are so aggressive toward each other. He is the reason why she has come here, to the University of Toronto. They are young and ambitious, and both are passionate about their research.

Tacked up on the wall is a faded clipping of an opinion piece from the

New York Times that her husband wrote. I vaguely recognize the article. It's about how different political groups have trouble communicating with each other because they prioritize different types of value systems and experience different types of "awe," to use Dr. Stellar's expression.

One group prioritizes human freedom, another equality. A third prioritizes the environment, and a fourth, the correct view of women's role in society. How can they talk to each other? And what role does the ideal play for the groups themselves?

I ask Dr. Stellar about her thinking around this research and its relationship to her own work.

"What does it mean that different groups experience awe in different ways?"

"A group that shares certain political ideals that they experience as greater than themselves can feel the greatest sense of community precisely through their shared awe. When we have a group that does something good for others in society, they also feel shared awe, and therefore community, in the group. The magic of these activist groups arises through their shared ideals, and that sense of community stabilizes the group and thereby the society."

In other words, awe binds people together and makes them feel good.

"But what about other types of awe?" I wonder. "Awe isn't just about political or ideological activism, is it?"

"No, then you have the other type of awe," Dr. Stellar says. "That all-encompassing feeling of nature. Or when you look at really beautiful art, or listen to music that fills your heart with emotion. Or when you feel that you encounter a good and present god."

"If the first type of awe, group awe, functions to stabilize groups and societies, then what is the function of the other type, the more individual one?" I wonder.

"We really don't know that," she says.

Let's take a closer look at the kind of awe that's experienced by individuals—for example, what you feel when you're out in nature and look at its beauty, greatness, and shifting character. The term "nature" is a broad one, after all. The waves of the sea, a little bubbling brook, a lush garden, and a quiet autumn forest full of mushrooms—they're all nature, but very different types of nature. Do all these nature types give equal results for inflammation?

"We saw some things that stand out," Dr. Stellar says. "There are two types of nature experiences that have a very strong anti-inflammatory effect."

The first type of "awe-inspiring nature" was large landscapes. Dr. Stellar takes examples from the magnificent landscapes of the North American continent.

"When you look out over a large-scale, open place, like the Grand Canyon, for example—that's big. Or when we took our subjects to the top floor here at the university and had them look out over Toronto. Views that give a bigger feeling of the world. I'd like to call it a kind of participation in something that's bigger than yourself."

Great landscapes present themselves in different ways for different people in different parts of the world, but they can all yield the same results, according to Dr. Stellar.

The other type of nature experience that triggers awe is more universal: sunsets.

"Sunsets—any kind of sunset?"

"There has to be a lot of color," Dr. Stellar laughs. "Sunsets create plenty of awe."

The sight of a sunset, a magnificent one in gold, purple, pink, and bright yellow—this is something that makes us stop and feel awe which decreases our inflammation.

But not everyone is able to live next to the Grand Canyon, the seashore, or on the top of a mountain with fantastic sunsets in sight. What about average city-dwellers, who spend most of their time surrounded by concrete, bus lanes, and traffic?

"The city can be very threatening for human beings," Dr. Stellar speculates. "I think that people in cities have to actively find their park, find their tree, seek out the nature that inspires awe, even in a city."

Dr. Stellar has become much more conscious of her own feelings of awe. What she used to consider pure recreation now feels like something important, a conscious choice. One of her favorite activities is visiting a museum.

"Tell me about beauty and art," I say.

"What you think is beautiful is also the thing that makes you feel awe," says Dr. Stellar.

I think about things that are experienced as beautiful. The dreamlike scenes and shimmering colors of the Impressionists captivate most people, as with Monet's water lilies. But art can be so much more, like Andy Warhol's stylized Marilyn Monroe portraits; Cindy Sherman's photo art, with its jabs at society; or tall, slender Giacometti statues in mid-stride.

"Are they all the same? Can you get as much of an anti-inflammatory effect from modern art as from the old Impressionists with their mild colors?"

"Certainly," says Dr. Stellar. "We've seen the same effects everywhere, from Impressionists and flowers to very modernist paintings and styles. We even had a person who was awed by architecture, everything from ultra-modern Bauhaus buildings to antique pyramids."

The response to images is instinctive among human beings and is linked to vision by age-old mechanisms, with connections deep inside our reptilian brain. The same is true for hearing. It doesn't surprise me that Dr. Stellar begins talking about music next.

"Music is incredibly important for awe," she says.

"Is all music equally good?"

"We found that music that triggers a feeling of great awe often consists of the sounds of many different instruments, like symphonies."

"And when in the music do you feel the most awe?"

"The sight of a sunset, a magnificent one in gold, purple, pink, and bright yellow—this is something that makes us stop and feel awe, which decreases our inflammation."

"When you approach some kind of crescendo, a kind of great center where the music is at its most dramatic."

But music doesn't only involve listening; it can also involve creating, preferably in a group. I think about choir singing and a study I've just read about that was carried out by the Royal College of Music in Great Britain, in connection with a cancer care institute. The study examined how singing in a choir affected mood and inflammation. It was found that inflammation levels decreased in everyone, but particularly in those participants who also suffered from depression. Studies at Harvard and Yale have also shown that singing in a choir extends life, literally.

But now I want to return to God.

The Swedish author Louise Boije af Gennäs, a friend of mine, has said that "God is the Swedes' biggest taboo." Or, as another one of my friends said, "The Swedes have done away with God, and so will people in other places in the world, since most people want to gradually become like Swedes." Well, maybe it isn't quite that simple . . .

It's often said that Sweden, like many countries in the Western world, is a de-Christianized country and that people have replaced Sunday service with a visit to Ikea. There's much to be said for Ikea's good, cheap beds and bookshelves. However (sorry, all you Ikea employees), trying to push a cart through an Ikea warehouse, laden with an assemble-it-yourself table, a Billy bookshelf, two plastic garden chairs that were on sale, and a dusty yucca palm, is not awe in my book.

But maybe Swedes have replaced the church with something else, like nature, for example.

According to religion scholar David Tufnell at Södertorn University, most Swedes claim to believe in some kind of spirituality or life force. In other words, people look for a spiritual experience, even though they may not find it in today's Swedish church. Perhaps the churches are doing something wrong?

Ninety-eight percent of people on earth consider themselves to be religious. But a better word is probably "spiritual." Most of us have a collective need for something—but what is that something?

"Oh, we human beings need to have spiritual experiences together," Dr. Stellar says.

The question then is, What produced this need for spirituality?

In May 2016, a study was published in the respected scientific journal *JAMA*, where scientists from Harvard showed that women who go to church have a dramatically smaller risk of dying early. They had studied 75,000 women over twenty years in this massive study.

Okay, says the skeptical cynic. But wait a minute—women who go to church, aren't they a little different? Probably nonsmoking, non-drug-using drinkers of elderflower juice, who spend Sunday evening with praise singing and a bike ride around the block, wearing a hand-knitted sweater? The interesting thing is that the scientists adjusted the study to account for differences in alcohol and drug use, exercise habits, mental health status, ethnic background, diet, and so forth. The only thing left as an explanatory factor was going to church.

The Harvard professor who directed the study, Tyler VanderWeele, said in an interview that "church visits may be a powerful and under-estimated cause of good health." The researchers noted that it made no difference which church the subjects attended, and that even the women who went to church only sporadically had a 13 percent lower mortality rate. The researchers had a harder time seeing an effect on men, but on the other hand, men weren't specifically being studied.

Professor VanderWeele's interpretation of the results is that the practice of going to church, in itself, creates strong social networks. Several other studies have demonstrated the same type of connection between good health and attending church. At Duke University, it's been shown that religiously active people have lower blood pressure and less mental illness. Another study, published in *Southern Medical Journal*, shows that people who attend church once a week or more go to the hospital less often and become bedridden less often.

Might our spiritual needs in fact have biological roots? Or perhaps in their search for divinity there are a number of other benefits that human beings can gain along the way—benefits having to do with ecstasy, transcendence, and the ability to build shared myths? Could it be the case that nature rewards human beings when they are spiritually active, by decreasing their inflammation? Do we simply need to experience spirituality and ecstasy once in a while?

"Spirituality is a communal act, to be able to bow down before something larger, together. We need it as a group of human beings," says Dr. Stellar.

Now we've arrived at the topic of God, or rather, the divine. It's a difficult subject, which I now find myself discussing often with my friend who is struggling with her illness.

"What about God? I would like to believe in something, but how does it work?"

That question, from her, just then, deserves a real answer here.

I have to reflect about it, since I don't want to force my beliefs on anyone. Besides, it's a private feeling within me that is hard to put into words.

"Hmm . . . I think of God as a collective name for the great universal energy, the absolute love and light."

"What do you mean?" my friend asks.

"God is the name we give to *the other*."

I sound like a greater theologist than I am, since I've borrowed the last answer from a former archbishop.

But if you believe, as I do, that there is something more than just the atoms that exist in everything, then you believe that there's something outside of us. That is the other thing that we call God, in other words, something incredibly huge that we might just as well call the universal life force and in other cultures is called Yahweh, Allah, Buddha, or Brahma. That's what my belief looks like.

"But, the church is so weird," my friend says.

Yes, then there's the church.

It is human beings, with all their strengths and weaknesses, with their will to build frameworks, to arrange and control things, who have built this church. Maybe that's why the church becomes as magnificent and as pathetic as human beings themselves are; that is, a house for all the people who are searching and fumbling around in spiritual twilight. Through the centuries, these church spaces have been spaces for rest, beauty, healing, and education and have welcomed people in their sorrow and vulnerability. But they have also been spaces for anxiety, high-handedness, and hypocrisy. Not everyone feels comfortable in this space, and I can understand that. But I look beyond these problems.

"I don't really understand," my friend says. "But I'd like to understand."

I barely understand it myself.

In my journey, I found a future clergyman to talk to. It wasn't easy to talk about spirituality; I barely had words for it. But I needed help finding my way.

This man was different. He was someone I knew as a casual acquaintance, who had changed his life path and made the radical decision to convert to Catholicism and became a Jesuit priest. He lived in a monastery in Wimbledon outside of London, and I visited him there.

We began talking about God, and I groped for words.

"Sometimes God is very close. But then I get lost."

"How is that?"

"I feel like a lost little monkey running around in a room, not knowing which direction to go. I feel like all the things that are urgent hide the things that are really important."

"And where is God then?" the man asked.

"He's standing outside the door, knocking. But I don't let him in."

"What would happen if you opened the door?"

Now I'm investigating what happens if I open the door—as in going to church more regularly with other people.

My friend wonders, "Why do you go there? What do you get out of it?"

I try to give her an analogy that might not be entirely appropriate.

"You can't sit on your butt and hope to get into better shape. You have to go to the gym or out on the track and train your muscles. I think of church in the same way. I can't sit still and hope to feel awe and be able to connect with *the great life force*. I have to go to church and *do* things."

"Like what?"

"Like pray, take communion with the others, participate in the music, and so on."

"So the church is your spiritual gym?"

"Well, maybe in a way . . ."

"And do you get stronger?"

"I think so, but my weakness is welcome too."

In any case, it gives me great peace and a new direction, a deeper connection to life itself, and a budding telephone line to the divine. Maybe that's partly because of God, and partly because of an anti-inflammatory effect that I get as an extra bonus?

Back to my conversation with Dr. Stellar.

"The feeling that God is great is important," she tells me. "It's important for creating a positive group feeling."

"Do all forms of belief lead to less inflammation? And can it be any kind of god?" I wonder.

"It has to be a good, benevolent, sympathetic god. It can't be a strict or judgmental god. That kind of god creates a negative awe instead."

And if I look in the dictionary, I see that *awe* can also be translated as "terror" or "fear." Maybe the feelings are neighbors?

"When you feel that you're a victim of something bad," says Dr. Stellar, "when you feel that you're very small, and the bad things are so much bigger and stronger, it doesn't give any positive biological effects at all. It could be an evil spirit, or a tornado outside your window. Or when subjects look at old movies where Nazis are worshipping Hitler. That's negative awe. That's terror."

"You're making me think of sects and cults," I reflect. "Demagogues and inspiring leaders who create a feeling of awe but also isolate their followers."

We talk about Jim Jones, the cult leader who made a whole group of religious followers commit mass suicide in the jungle in Guyana.

"Can evil leaders manipulate people's capacity for awe and take advantage of something that's actually a good force in people, for their own ends?"

"I'm sure that's the case," Dr. Stellar says thoughtfully.

We are quiet for a while. She continues:

"We also need to reflect on whether awe creates a sense of 'us' among us, and if in that case it also creates a 'them.'"

"What do you mean?"

"If we, in this group, feel awed by something, on this occasion, but that other group doesn't, then we don't belong together, since we feel awed by different things. We don't know yet how this is transmitted chemically, but a great deal of work right now is focused on the hormone oxytocin."

And there it turns up: oxytocin. Interesting.

In the 1990s, I wrote a book about the body and soul of new mothers, and research on oxytocin was new then. Oxytocin is an intricate hormone that has followed people since way back in evolution, long before people became human beings. It stems all the way back from when we became mammals.

Oxytocin is many things. First, most concretely, it's a hormone that helps women give birth and nurse babies. The uterus is a remarkable muscle in that it actually rests all through life, except in connection with pregnancy and childbirth, when it is suddenly expected to perform a marathon. This is the substance that helps the uterus carry out its enormous task of giving birth to a human baby. It's also the hormone of nursing. When the baby suckles at its mother's breast, oxytocin is produced, which helps to express milk from the breast.

But the function of oxytocin has turned out to be much more sophisticated than this. The substance also creates feelings of love and intimacy, and forms bonds between people. It has even been found to be wound healing and anti-inflammatory.

Why am I not surprised?

And now Dr. Stellar tells me about the next step in her research. The very things that form bonds between certain people might also cause a *non-bond* with other people. Maybe you could call it an "us and them hormone"?

"The research that's being done shows that this is very complex. It's one of the oldest hormones in mammals, and we know that its effects are related to motherhood, nursing, and creating strong bonds. But it isn't just a fuzzy love hormone that creates trust in absolutely everyone. Maybe there's something that also excludes others?"

I take notes and ask her to say more about the complexity.

"We're thinking a lot about social inheritance and inflammation. We see that generally, in people who experience themselves as living in a socially vulnerable environment, negative feelings seem to more easily trigger higher levels of inflammation markers."

She gives me an example. If you ask people to solve a difficult math problem in front of a large audience, something that stresses most people and therefore triggers negative feelings, those people who come from a socially vulnerable environment will have higher levels of inflammation markers in their saliva than those who come from more stable environments.

"What can you do to counteract this feeling?" I wonder.

"Research shows that a secure person in the audience, someone the person feels is on their side, can decrease inflammation in those who are more vulnerable."

Inflammation once again. But there's even more complexity to this— and it has to do with cultural differences.

Dr. Stellar's team did studies in China that paralleled the ones in the United States and found great differences in how awe is experienced.

"In China, awe is more related to negative feelings," she says.

"Could it be related to language?" I wonder. After all, the Chinese have a different way of building up their language.

"We had a hard time even translating the word. All the synonyms we could find were linked to social status in some way, with the approximate meaning 'I worship this person who is on a higher rung than I am.' Like something they might express about a leader or a boss."

We discuss why this might be so.

"These are speculations," Dr. Stellar says, "but we could see that their awe was almost always linked to fear, to powerlessness, to some form of control. We did not get the same results at all."

"Did you find any explanation?"

"No, we don't know why. The American feeling of awe is much more optimistic and positive than the Chinese one."

These are large questions, and there is much research to be done. But in order for us to fully believe what Dr. Stellar tells us, I still have to ask:

"Is it definitely true that awe directly decreases inflammation, or is it just a correlation that takes place via some other system?"

This is what I'm thinking. In the 1960s, there were alarming reports about women getting skin cancer from the Pill. After a while, people realized that it wasn't so. Women who took the Pill were somewhat younger, often single, and this group was among the first to begin going on charter trips, where they were sunbathing much more than they could do at home. There was a correlation—these were the same women. But there wasn't a direct link, because it wasn't the Pill itself that caused skin cancer, but sun exposure. It's something like this I'm getting at.

"That's a valid question. We need to look at it more closely, and we'll need a much deeper understanding of the biological mechanisms at work."

The sun has begun to set outside the window, and the traffic is thickening on the streets of Toronto. There's the sound of a police siren farther away.

"And what's your awe?" I ask her. "What creates awe in you, as an awe researcher?"

"I love traveling," she says, and her eyes light up. "So it was a great thing for me to see the pyramids in Egypt. Also the Taj Mahal in India, and Angkor Wat in Cambodia."

Angkor Wat is a temple built by Cambodia's ancient rulers, the Khmer, in the twelfth century. You drive and drive through jungles and across desolate rice fields. Suddenly, like in an Indiana Jones movie, the Hindu temple rises up from the jungle, with elegant, sensual goddesses dancing on the walls in the sunset, and with turrets and towers; abandoned by time, it is still inhabited by myth and stories. I have seen it, and was struck by the same sense of hidden history and romantic magic.

"It's an absolutely fantastic temple, so steeped in history, and it just emerges out of the jungle. It has everything," says Dr. Stellar.

But you can't be in the Cambodian jungle every day. What does her everyday life look like?

Being able to exercise outdoors is
magic and motion rolled into one.

"I walk to and from work. The fall is fantastic here in Canada, with beautiful maples in vibrant red. That's everyday awe for me."

The visit makes me think. Here we have a trail that's all about experience, insight, awe, and spirituality, and it is directly linked to lower inflammation markers. It isn't completely covered by formal religion, even if religion and a belief in God seem to be central elements. You can see that in both Jennifer Stellar's research and empirical data from the Blue Zones.

Since I'm already on the other side of the Atlantic, my trainer Rita

DR. STELLAR'S AWE-INSPIRING LIST

1. **Nature.** Look at pretty sunsets—the more colors and the more spectacular, the better. Big, open landscapes also top the list. Go to the Pacific Coast, walk in the mountains, or boat in an archipelago.

2. **Art.** Go to the National Museum, go on a gallery tour, whether it's Monet's Japanese bridge or German Bauhaus architects; go to the ballet, open yourself to artistry that makes you think and feel different.

3. **Music.** The grander, the better. Many instruments and sweeping crescendos, whether it's Grieg or Queen. If it gives you shivers, you're in the right place.

4. **Meditation.** Slow down and listen. Find a guru or an app, follow Deepak Chopra. Just find something that works that you can do every day, and that gives you rest in a waking state and a little everyday ecstasy.

drives to Toronto from her hometown of London, in Ontario. We meet and discuss everything that's happened since last time.

And we talk about awe, in its modern form.

"What do your clients do to feel awe?"

"It's an important piece," she says. "Daily gratitude, meditation, mindfulness—there are many techniques."

"Do you see a difference when people use these techniques regularly?"

"It might make all the difference," she says. "But sometimes you have to take big steps."

"Big steps? Like what?"

"Maybe you should try going to a bliss course."

Bliss course?

5. **Spirituality and religion.** Look for symbols and contexts that touch you and make you feel that life is bigger than just here and now. A moment in church or a prayer, a visit to a temple or lighting a daily candle, reading something from a spiritual book, an app with spiritual meditations, spiritual music, drums . . . Where does your spirituality live?

6. **Doing something for others.** Join Amnesty and protest North Korea's prison camps. Join Greenpeace and chain yourself to an old oak tree. Join the Red Cross and fundraise outside the grocery story on Saturday morning. Get involved in something that's bigger than you, makes the world a little better, and decreases your own inflammation at the same time. Win-win.

7. **Watch sports with others.** Sorry, not in front of the TV, but live. We are talking about the collective ecstasy you feel with others who like the same team, at an arena. Go to a soccer game, basketball game, a diving meet, or something else where people are engaged and you're engaged.

*The greatest gift
you can give another person
is your own happiness.*

—*Esther Hicks*

11. BLISS

I'm lying on the floor in a big room with hundreds of women that I've never met before. It's early morning on the Pacific Coast, just south of Los Angeles. A shaman is playing drums in the background.

How in the world did I end up here?

The night before, I was sitting in the never-ending traffic around Los Angeles, along with a lanky truck driver with a side gig as an Uber driver. He normally drives the interstate route, the toughest route an American driver can take. You drive between the fifty states of the USA, are gone for weeks at a time, spend nights in the little space behind the driver's cabin, and live on hamburgers and Red Bull. Apparently, Colorado is the most beautiful state, but many states, "none named and none forgotten," are, quite bluntly, crap, says my newfound friend from his seat at the wheel. He's wearing a cap, T-shirt, and jeans.

"Where are you going?" he asks.

"I'm going to a bliss workshop."

The guy casts a suspicious look in the rearview mirror.

"What?"

"It's a weekend course for women—my personal trainer recommended it."

"Oh. What do you do there?" the driver wonders.

I tell him about my book and that I'm looking for new ways to lead an anti-inflammatory lifestyle. And that I suspect that women today are looking for awe in these kinds of new ways, like workshops about lifestyle and inner joy.

"Hmm, that's new to me," he says. "But maybe it's something for you women?"

Then he begins telling me about his own way to get happy, about his BMX bike, how he's been learning new tricks on it ever since he was a boy. Bunny jumps, 360-degree turns, jumps over big obstacles.

"We used to build bumps in our nice forest back home, bump after bump. We'd go completely wild, just jumping!"

It's like Jennifer Stellar says: awe can take many forms, and this man has found his awe.

But for the women who have paid $500 to take this workshop, it's a different journey that beckons. What is that journey, and how is it connected to the anti-inflammatory tricks that I'm now beginning to see more and more clearly? The right food, the right exercise, meditation, spirituality, nature, breathtaking experiences . . .

This is what I want to examine.

That's why I'm now lying here on the floor, packed in like a sardine among all these other women. That's why I'm putting up with this course.

That's what it feels like—as if I don't fit in at all.

I've already noted that I'm older than most of them, as well as being from Europe. The girls in the room next to me come out into the hallway of the hotel early in the morning wearing full makeup, complete with bronzing powder and false eyelashes. I listen to them talk in the elevator. They run some kind of company in Arizona, something having to do with photography and fitness, and already at six in the morning they've had time to send away two Instagram photos to market their new business idea—a persistence that deserves my admiration in every way.

We're staying at a classic American middle-class hotel with a Starbucks in the lobby. Outside, on the coast side of the hotel, California's tall, slim palm trees sway against the sky, where the sun is just rising. But inside the yoga studio, it's all dark.

We're working on our yoga mats, close to each other. It's crowded, and

I can feel warmth from my neighbor Binu's skin and hear her breathing. Binu is the only one I know from before. She's Indian/Canadian and is also training with Rita. Binu is an entrepreneur—open, smart, and generally on top of things.

I'm not on top of anything right now.

Our yoga instructor begins with a story about how she left her job as a high-powered consultant in business for a life as a yoga teacher in Costa Rica, the new travel destination for global yogis. She went through a crisis just as she was moving, and now she's talking about sisterhood and inner healing, as she helps us to breathe slowly through the vinyasas. She isn't ethereal and quiet, like European yoga teachers. She talks the whole time, and plays loud music—everything from rock music to nature sounds, R & B, and ballads. There's a lot of sound and noise in this lesson.

When we go from Downward Dog to High Lunge, she asks us to sway our hips and love ourselves. We stand there for a long time, swaying from side to side. She urges us to sink deeper into our "self-loving," twist our hips, sway like a figure eight, and think about where it feels best. The yogi talks to us about the inner goddess who lives in every woman.

Okay, I'll sway. But as for my inner goddess, I haven't had much time for her lately, I must admit.

"Hey, are you in there?" I call internally.

There's no answer.

❧

The inspiration and arranger behind the course is Lori Harder. She started her career as a fitness model in an American small town, and after winning won the Miss America title, she became interested in taking a bigger step, toward the wholeness of body and soul, which is the mantra for her enormous Facebook group. Naturally she now lives in California, which has been the ultimate fitness frontier ever since Jane Fonda's time.

There's something in the light here, in the meeting of East and West, land and sea, that promotes new thinking—not only within fitness and inner development but also in the explosion of everything digital.

The course is a California blend, combining digital media, entrepreneurship, management, and fitness; but above all, it's a journey toward bliss and inner joy, through empowerment, meditation, yoga, and the art of making peace with old mental baggage that holds us back. Lori Harder is the queen of Instagram and arranges courses every year that have become more and more popular.

This workshop is supposed to support the participants' journey toward perfect harmony in body and soul, total bliss, fulfillment, and peace. I wonder if it's a stage of total anti-inflammation, where body and soul, heart and brain can work at their maximum capacity on a kind of optimal original level. Is this what the women here are looking for?

Three hundred fifty women have come to the course from Arizona, Minnesota, Texas, New York, and, of course, California. They've come from Canada and the Bahamas, and a few from Europe and Russia. Most of them are hungry and fit, advanced users of social media, entrepreneurs, marketers, and sellers. But I see other types of women too, who must have quite different backgrounds.

A shy, overweight young woman with thick glasses, for example, stands awkwardly in a Triangle pose a few steps away from me in the room. She has a hard time reaching her calves with her hands. There are not only well-trained gymnasts here but other kinds of people as well. I see her over the course of the workshop; she keeps to herself.

We finish up the session, hands pressed together in front of the heart. Our yogi talks about how we can let go of fear and everything old and continue looking for our inner goddess. And now comes the shaman.

He creeps up onto the stage next to her, takes out a drum, and begins playing to the flowing, relaxing music and birdsong. He has long hair that swings in time to the rhythm. To the low voice of the yogi, many of the women sink down as if in a trance. I see several women around me begin to cry.

Something is resting. Something's waiting.

But not for me, of course. I'm just here to watch.

At breakfast, I observe how the women choose their food at the large buffet. Boiled eggs, scrambled eggs, smoked salmon. At the buffet there's a juicer, with bowls of cucumber, beets, carrots, spinach, celery, pineapple, strawberries, cantaloupe, and apple. I see that the women make juice, especially using vegetables. (They have what's called "a clean palate" in fitness circles, or a palate that's been readjusted so it doesn't crave sugar. Because they normally eat foods that have a lower sugar content, they are better able to taste small amounts of sugar in foods.)

A British man, who probably enjoys having an extra pint at the pub, judging from the tight shirt that's clinging to his stomach, is watching all these fit, young women squeeze their juice.

He turns to a woman in purple Lululemon workout clothes.

"How can you drink that bitter stuff? Don't you want to put some fruit in it to make it taste like something?"

"The less sugar, the less inflammation," she answers.

The knowledge is starting to spread, I note.

I take a quick shower in my room, where the air-conditioning overpowers the highest volume of my iPhone music. We're definitely in the United States, I think to myself. Then it's time for the high point of the workshop.

❧

We go into a conference room.

Loud R & B music is pumping out of the loudspeakers. In front of each chair, a large bag is waiting for us, from what may be the American continent's most successful sports brand. They make exercise wear for the new age but also market it as something more—a lifestyle and an attitude. On the bags are encouraging slogans that I have time to study since I'm a little early and also am sitting by myself. I read skeptically. *Breathe deeply.*

Thanks—that's always good. *How you look at life is a direct reflection of how much you love yourself.* Okay—banal. *Maximize your creativity and live in the present.* Easier said than done.

This is a goodie bag, an expected part of American self-actualization workshops, I gather. We dive into the bags. They contain organic almond butter in little plastic packages; organic cherries and beets, squeezed to make a puree, or *fruigee*, for on-the-go antioxidants. Muesli made of gluten-free oats, raw coconut oil, honey, almonds, rice protein, durra, raw cacao, dates; a caramel and almond bar with nuts and seeds, described as gluten free and with low GI; and a new type of nut butter that combines flaxseeds with Brazil nuts, sea salt, almonds, and raw cacao.

The American food industry has apparently begun to open its eyes to the anti-inflammatory mantra "Less gluten, less sugar, less lactose, good fats, lots of protein." The industry is now in full creative innovation mode, especially many smaller businesses on the west coast of the United States, which are a step ahead in this food transformation. New products appear that our forefathers wouldn't even have recognized as food. And it's the women here at the workshop who are the target group—what marketers call early adopters.

The music gets louder. At the rear of the stage, a movie screen appears. A classic black-and-white Hollywood countdown. Then the words roll out, big and flashing. The women clap their hands.

Possibilities. Intention. Sisterhood. Examine. Support. Your time is now. You are connected to something much greater.

Four dancers come in and dance, rocking to the R & B music. Then it's time for our bliss queen, Lori Harder herself.

She steps up onto the stage, to the excitement of the crowd. She's incredibly beautiful and fit, with long blond hair and bare shoulders, and is wearing a long, wine-red pantsuit with large white flowers. She tells us that she felt nervous about getting up onstage.

"But," she asks rhetorically, "how do you handle a giant, scary wave?"

People call out the answer. The guru nods.

"You dive right through. And that's what I'm doing here with you now."

She talks about the anxiety symptoms she had as a little girl, how anxiety and fear followed her through life, and how she's learned to welcome anxiety and hold it like a newborn baby. She demonstrates holding the little child in her arms.

"Then I tell my anxiety, 'There you are, you always come when I'm going to do something good.'"

She asks rhetorically how you can climb out of a feeling of anxiety to reach self-confidence and happiness, and when she answers her own question, the answer is completely unexpected for me.

"You don't have self-confidence. You *make* self-confidence," she says.

The crowd cheers. This is as far as you can get from the Scandinavian law of Jante, which tells you not to be ambitious or do anything out of the ordinary. Here, there's a total trust in the idea that we human beings are capable of all kinds of change, worth every success, that we can achieve everything in the external world if our inner world first raises itself to the right level of consciousness. I begin to debate with her internally, formulating silent but good counterarguments for why it may not be that simple—but I barely have time to do that before she takes the next step.

"Since you can only create change in yourself from a higher stage, I'm going to lead you through a meditation."

The light is dimmed. New Age music streams out of the loudspeakers. We close our eyes and connect our thumbs and forefingers.

We are invited to meet ourselves, free of anxiety, and open up to our higher selves. A beautiful image appears. My higher self comes to me like a beautiful and kind being, perhaps a librarian from my childhood, someone with endless wisdom, surrounded by a blinding light. I realize that she's available to me at every moment and that she is a welcome break from my everyday self.

Lori Harder quotes Einstein.

"Involvement and creativity are more important than knowledge. For knowledge is limited to what we already know and understand. But our capacity to imagine encompasses the whole world, and everything that will ever need to be understood."

As I sit there listening, I realize that my earlier feeling that this is a much too simplified view of people isn't really accurate. I begin to see that Lori Harder is a psychologically developed person with a feeling for the complexity in all of us. We do writing exercises, and have a group therapy session where we shout out new mantras that reject all the old baggage we drag around with us. A woman who has felt stupid ever since her school days calls out that she's smart.

"I am SMAAAAART!" she shouts as she falls back into the arms of three other women.

Another one shouts that she is beautiful, as opposed to her old feelings of being ugly. Many of the women shout that they feel powerful.

All of this is good and commendable. Still, the whole event feels alienating. Or maybe I, with my new knowledge, can see that I choose to alienate myself.

Because it just feels too American and culturally foreign to me.

Because I hardly know anyone at the course and don't really feel at home at this giant women's meeting.

Because I think there are too many slogans.

But I also realize that all of this lives inside me.

In the evening, my jet lag hits. I skip the disco with three hundred fifty other women in makeup, as well as the opportunity to pose for a photo of myself with the beautiful guru.

The next morning, I seriously consider dropping out and leaving.

I've understood what this is all about, I think. The lesson is clear. This is what awe looks like in California. But now I'm struck by something. This type of thought is part of a pattern that I have. When something is new and informative but rubs me the wrong way, I want to pull away.

Okay. What would happen if right now, I try something different?

The possibility intrigues me. I take my spidery I'm-leaving-now feeling in my arms like a newborn baby and softly speak to it:

"Hey you, you little 'I'm leaving now' feeling. Why do you always turn up when I expose myself to unpleasant, but educational, new situations?"

I end up staying. This morning, once again, begins with our yogi.

And now I see how she gives all of herself. We work on opening our hearts. We lie down in a Rockstar pose. It is intensely uncomfortable, and my whole hip aches.

"Ah, a little discomfort," the yogi jokes with us. "Welcome the discomfort. Let it come and go. You can stay with something even if it's uncomfortable."

I almost blush. Can she read my thoughts this morning?

And so a new day begins, with writing exercises and talking in pairs. Now I'm my skeptical self again. But this will be disrupted in the most unexpected way.

Today Lori Harder starts talking about hugs and how much she likes to hug people, with long, warm hugs. Then she urges us to go around the room and collect long hugs—among three hundred fifty women, most of whom are total strangers. We are to get as many as possible.

The light is turned off. This is embarrassing and uncomfortable, even for me, a mother of four who's used to hugs.

Hugs have been shown to free up oxytocin, the hormone that creates comfort and strengthens human bonds, which I discussed with Dr. Stellar in Toronto earlier. You have to hold the hug for at least twenty to forty seconds in order to get the hormone going; in other words, beyond the limit of where it begins to feel embarrassing.

So, we begin hugging.

I get shy hugs, eager hugs, nice hugs, warm, heavy hugs. I feel shy at first but soon begin to see how much warmth and light and goodwill exists in these women. I begin to see how beautiful they are, each one in her own way.

Then she turns up—the lonely and heavy woman who struggled with her Triangle pose the first day. She stands a little apart in a row of benches, with her head hanging and her hair over her face as if to hide herself. She seems shy and hesitant. I go up to her, and we hug. She presses herself against me, all warm, and starts crying uncontrollably. She collapses completely, and I have to hold her up. I don't know who she is, or what's happened, or what planet she comes from, but she almost sinks into me. And I realize something.

I realize that I need to stand here exactly as long as she needs me to, and just hold her. I think that no one has hugged her in a very long time. I wonder if all this, in some cosmic way, was the real meaning of the workshop: that this lonely young woman would be held and that I would be able to give her security and closeness right now; and that my "I'm leaving now" feeling from the morning would be overcome and I would learn this lesson.

I realize something else: I get just as much truth and warmth out of this long, sincere hug from this woman with so much sadness in her, and this also gives me a chance to release my old sorrows and feelings of loneliness.

This experience creates the strongest feeling of awe in me.

After this, I see both the course and the world in a new way. I'm small in relation to this large thing called life. I feel a great sense of awe, which corresponds exactly to Jennifer Stellar's model of awe, and it feels fantastic. I can see my IL-6 falling, as if I were a subject in Dr. Stellar's study group.

I decide to become one with the workshop and let go of my skeptical superego. It's a wise friend sometimes, but now I ask this friend to lie down in the room and rest for the remainder of the workshop. Or maybe come visit now and then, when I so decide.

We sit in groups and write little, beautiful messages to each other in gold text on candles that will glow when we get home and light them.

I make my own list of things that give me bliss. I write, in no particular order: being out in nature, singing in a choir, spirituality, reading good poetry, being in a temple, lifting weights, music (jazz, opera), being close to family and the children, working with things I'm passionate about, yoga, art, deep and intimate friendship. This is my awe just then.

We must answer the question: How can you do more of this in your life?

I buy everything, even the Bliss shirt, a grayish sweatshirt with the course name printed in gold letters across the front. This will turn out to be meaningful in a few weeks.

Until then, I take these lessons with me:

- We have access to our higher selves every day and must connect to the divine wherever we find it. There we will find power. Americans are good at dealing with spirituality, which is allowed to exist in all kinds of shapes and forms.
- Never underestimate a hug—a long hug. Dare to seek out the unhugged and receive—give and receive.
- Stop when you feel uncomfortable and ask yourself why the discomfort is happening right now. Maybe there's a message, something that waits for you if you dare to stay?
- Stop sending in what Lori cleverly calls "the representative," or the version of yourself that you think other people want to meet—the perfect, well-groomed, and articulate version. It makes you tense and blocks close relationships.
- You can welcome pain and negative thoughts, become friends with them, work through them, and get rid of them.

And when I end up next to a quiet woman with long pretty hair and longer eyelashes than I've ever seen before, we tell each other about who we're going look up when the workshop is over, to ask for the help that we need. I decide to look up a person who will help me get rid of some unpleasant baggage that I'm carrying, an old burden of sorrow and guilt that sometimes paralyzes and diminishes me.

My partner tells me that she feels tense and stressed by all the obligations, by other people's perfect lives that she sees on social media and compares to her own, somewhat messy life. She wants to find a balance, an inner security, and decides to seek out someone who actively meditates every day and can tell her what it's like.

Strangely enough, I know exactly whom she should talk to.

"Hugs have been shown to free up oxytocin, the hormone that creates comfort and strengthens human bonds."

*Meditation is a long journey,
not a single insight or even
several insights. It gets more
and more profound as the days,
months, and years pass.
Keep reading and thinking
and meditating.*

—Dalai Lama

12. PEACE

The person she's looking for might just be me. And I tell her about my journey.

I can see it in front of me, the hall of the Stockholm apartment where we used to meet. It had a very unusual, deep pinkish-red color that enclosed us like a uterus.

We were a motley crew. There was a tired woman who worked at the post office and went to work at five o'clock every morning; a journalist, my close friend who had almost forced me to come; a high school teacher; a very tall entrepreneur with a beard; and a nurse.

Why were we all gathered here? We were hoping for less stress, more peace. Maybe enlightenment in some form—I don't know what. Liberation from negative thoughts.

And so we gathered in the room off the pinkish-red hallway, bewildered and expectant. We were to be initiated. It sounded strange, but also special. We were carrying flowers and fruit, which we were going to give the teacher, who was waiting in another room. We giggled at the sight of our bruised bananas in simple plastic bags from the co-op grocery store around the corner.

When I entered the room, I found the teacher, a middle-aged Swedish woman, standing at a table. She was wearing a red sari wrapped around her waist. On the table was a silk tablecloth and a copper plate holding incense, which burned slowly with a sweet and smoky smell. There was a photograph on the table of an Indian man with a beard. The beard was gray and wild, almost frightening, but the man in the picture was laughing. A red necklace hung around the frame.

"You can lay the fruit and the flowers on the table," said the teacher in a broad Stockholm dialect that contrasted sharply with her sari.

The teacher closed her eyes. Her Stockholm dialect morphed into a foreign language. Could this be Sanskrit, or Hindi? She was singing a monotonous, ritualistic song. I wondered why I was even sitting in this strange room. The whole thing was surreal, even more so in the 1980s, when the world was still quite closed and different cultures still upheld rigid barriers.

A friend and colleague had awakened a question inside me.

"Why do you get so worked up?" she asked once when I was really going crazy.

"I really care about my work and I want to produce results," I answered. "But then I work so hard that I deplete myself—my batteries run out."

"Just remember, there's only one person who can help you handle your stress," she said.

"Who?"

"You," she said.

"Me?"

"No one else is going to turn up and help you with this."

"Hmm, maybe you're right."

"What are you going to do?"

Yes, what could I do to deal with my stress?

❧

I was working as program director for the first so-called infotainment program on TV. We mixed news with entertainment—one minute, an update on Israeli settlement politics, and the next, an international pop star singing a tune from the top one hundred.

The year was 1989, and it was the first time serious news reporting and performing artists were mixed that way on Swedish TV, even though the genre was already commonplace in the United States. We were seen as a symbol of degeneration of serious TV news and became the target of small but steady attacks in the media.

The program in itself also had many stressful aspects. In addition to a fast production pace and daily broadcasts, I was pregnant for the first time, which took almost all my energy.

One day, the Spanish crooner Julio Iglesias visited the studio. He was adamant about showing only one side of his face to the camera. Before our prerecorded interview, we had to rehearse forever so that the camera could capture the right angle. Julio moved like an Egyptian temple decoration as he glided along the walls with his face in profile toward the camera. All the while, a reporter from one of the evening papers was following the spectacle. The next day, there was a four-page spread with pictures of me chewing gum and blowing bubbles as I waited for Julio to move with his right side toward the camera.

My friend Annika also worked at the program. She meditated every day, and I saw her smoothed-out face and radiant air when she came out of meditation.

"I want to try it," I said.

I later changed my mind, but Annika forced me to go anyway, for which I'm eternally grateful.

That's how, a little later, I found myself in that apartment. My teacher slowly intoned in the foreign language, and gradually a word appeared, like an animal emerging from the fog. The word was repeated again and again. It was beautiful. The teacher leaned forward, almost whispering.

"This is your mantra. Only your mantra. You must never reveal it to anyone."

We learned that transcendental meditation, or TM, was one of the many meditation techniques that found their way from Asia to Europe in the 1960s. TM had been developed by Maharishi Mahesh Yogi, the laughing man in the photo.

The goal of the meditation form was originally a kind of enlightenment, as manifested through Buddha himself. When Buddha was only nine years old, he meditated for the first time under a tall tree on the outskirts of an Indian village. He observed the people in the village, the birds in the tree, the worms in the ground, the water buffalo in the fields, and how all of them were interrelated. All creatures were different, but at the same time they were all victims of life's brutality, united in their attempts to avoid suffering.

Back then he wasn't even Buddha yet, but Siddhartha, a spoiled Indian prince who was just about to begin his journey toward becoming a buddha, or enlightened one. And on this journey, meditation would be the key.

It sounded large and amazing. But TM had a simple framework. I was supposed to sit down on my behind twice a day, close my eyes, and not think about anything in particular.

Well, except for the mantra, which I understood more or less as follows:

A mantra is a kind of sound image with roots in Sanskrit, the ancient Hindu language. Within TM the idea is—and now it gets fuzzy and I can't quite follow—that the mantra imitates the natural sounds that have existed since the beginning of creation and thereby have the ability to reach deep down to the cellular level and heal body and soul.

The mantra in itself doesn't mean anything. It's a series of letters that form dynamic sounds. The most well-known mantra, the universal mantra, is Om, which resembles our own breathing.

The function of the mantra is to create a gathering point for the thoughts, a central station of calm that we can come back to anytime from the erratic wanderings of the mind. TM describes the human mind as being full of changing and flitting thoughts, like having a tree full of monkeys inside. These monkeys chatter constantly in our heads, about stress, conflicts, and forgotten laundry schedules.

Then they're up and running, swinging merrily around in the tree among upset feelings, old injustices, simmering discontent, inferiority complexes, lost drives, tough sorrows, and everyday anxiety. They make us feel exhausted and stressed and they diminish us. They can and should regularly be stilled with a strategic and systematic technique.

Meditation was just such a technique, according to the teacher.

Every time a thought turned up, I should just calmly return to my mantra, without judging the thought, said the teacher.

That's how we were supposed to begin meditating.

First there was this business of how you're supposed to sit. In the movies, I had seen that real meditators sat on the floor in the lotus position, with legs crossed and hands resting on the knees. The thumb and forefinger were touching and pointed upward, ready to collect energies.

No one in the group could sit like that. We were all too stiff in the hips and knees.

"It doesn't matter," said the teacher.

So our teacher let us sit in regular Ikea chairs, with feet on the floor and arms hanging by our sides.

We sat on our chairs, turned inward, as we worked with our mantras.

"No, don't work," said the teacher. "Just let the mantra come naturally. Don't judge your thoughts."

But the thoughts came anyway.

You don't have time to be here . . .

Breathe deeply and pick up the mantra.

My leg is tingling . . .

The mantra.

Why doesn't that woman at work like my new idea?

The mantra.

For every round of pulling the brain back to focus on the mantra, the inner voices began chattering a little bit less and grew somewhat less demanding.

After a while, I began to see my mantra as a stylish lady in a simple, light-colored dress, sort of like a Swedish St. Lucia or a dignified librarian, who glided in among my confused thoughts and told them to be quiet for at least twenty minutes, preferably longer. (This was the same librarian who came to me later, at the Bliss camp—I've always had the highest respect for this career that has drawn so many knowledgeable and impressive women.)

Soon I begin to see a pattern.

Thoughts were nothing more than clouds drifting by in a summer sky. They passed, and the mantra pushed them forward elegantly. Push . . . Push . . .

Finally, something else arose. Emptiness. And out of this—an inner light that I could only describe as peace, in the most magnificent meaning of the word.

Perfect joy, deep contentment, peace, and stillness.

I met a new side of myself, as if layers had been peeled away. Insecurity, defense mechanisms, stress, performance anxiety, the fear of failure or of not being liked—beneath all this lived a very kind, light, and happy being who wished others well and had no ego, who seemed to melt into the great wholeness of humanity.

"Could you say that this is my real self?" I asked somewhat naively. "Or maybe even that I don't have a self?"

My teacher looked at me with a kindly glint in her brown eyes.

"The spirit is pure joy," she said in English. "The soul is pure and light energy. That's what all of us look like inside when we're free of stress. This light, free point is point zero for all of us. But stress takes us away from point zero. Stress leads us to all of our bad actions, and away from our light."

"So this zero point, this light, joyful being in all of us, where we might even be some kind of common energy, is available to us if we meditate regularly?" I wondered.

"Yes, it's just that hard. And that simple," said my teacher.

That's how my life as a meditator began, and it has looked pretty much the same for more than twenty-five years now.

Every morning when the alarm rings, I go into the bathroom and brush my teeth, and then I sit down on my bed, simply and without fuss. Twenty minutes to meet the inner light and point zero and all of that. But this is actually an exaggeration, for there have been many periods in my life when it's been a struggle.

There have been weeks when I've forgotten, or when I haven't had time. When the children came, months would go by when I didn't have the peace of mind to sit by myself, quietly, because I was anxious about the children and always wanted to follow them with my gaze. This was especially true for the little one who used to poke pencils into electric outlets, the same one I once found after he'd disappeared, an eighteen-month-old baby running by himself straight into traffic, in a snowstorm, dressed only in a diaper. On certain nights, I would wake up seven or eight times to nurse or soothe unhappy children, and I was worn out and tired to the bone. Besides, every moment when I wasn't dealing with children or work had to be devoted to other things to keep the puzzle pieces of life together. There simply wasn't any time to sit and gaze at my inner self.

But something pulled me back, every time, back to the stillness and point zero of meditation. Now, after twenty-five years, it's gotten to the point where I long for my meditation, my bliss. It's as if it lights up my inner light in the morning, a light that lasts all day. It's my fix, my drug. Without meditation my day becomes grayer and harder.

But sometimes the mornings are hectic, or I'm traveling and my schedule is topsy-turvy. How will I make it work?

I've decided it's better to get it done than to do everything perfectly. I've learned to meditate à la carte—on buses, planes, in waiting rooms, at work in a corner, in hotel rooms, on trains, in cars, at friends' houses, on the bathroom floor, or in the children's beds after the goodnight story. I've meditated in sweatpants and pajamas, bikinis and party dresses, jeans

"I've learned to meditate à la carte."

and disco dresses, and dressed up as Santa Claus and Maria Callas. The only thing I've never been able to do is meditate outdoors or after drinking wine.

When I began the Rita program, meditation just followed along. It slipped in through the back door and settled down in my new bower.

A thought occurs to me: Do these things go together?

Over a twenty-four-hour period, the body moves in wavelike fashion between wakefulness and rest. One of the hormones that regulates the degree of wakefulness is cortisol. It's secreted by the adrenal glands and is kick-started partly by the body's own rhythms. The cortisol level is at its lowest late at night and early in the morning, just before waking up. Then it rises naturally, and we wake up and get going with our day.

Cortisol can also be kick-started by external stress. Think of that rush of energy and clear thought that you experience when you're late and run to catch a train that's just about to leave, or when you have to deal with a tough challenge. It's cortisol that responds to external stress and mobilizes the heart and muscles and thought apparatus to handle the problems.

This is roughly how it works:

- We are stressed by an external factor. Regardless of whether the stressor is being chased by a large animal, the thought of a child being hurt, or the worry that North Korea will blow up the earth, the response is the same.
- A cascade of hormones is set off, activating the adrenal glands, which begin to secrete cortisol.
- Cortisol sets up the body for a so-called fight-or-flight reaction, meaning that we are ready to escape or fight, depending on which is smarter. We might run away from a wild animal but fight for a threatened child. A large amount of sugar is released from the liver,

which has to supply the muscles with nutrients for the upcoming work. The sugar is broken down into another form of sugar in the blood, glucose, which is the smallest, most broken-down form of sugar, and easy for the muscles to use as nutrients.

- The cortisol prevents the insulin from storing glucose in the liver again and makes sure it's kept free in the blood.
- The heart begins to pump harder.

Normally we can now solve our problems—for example, make it to that train—and then the hormone levels return to normal.

But when you experience constant stress, you get new boosts of cortisol and also a kind of constant sugar boost, which in turn leads to the insulin resistance that Inger Björck talked about earlier—the one that is linked to the metabolic syndrome and, of course, inflammation. In other words, a little stress now and then is invigorating, but long-term stress can give rise to a condition of constant inflammation.

Imagine an existence with constant external stress. The phones ring with urgent demands; there's trouble at home; you drink acidic work coffee five times a day, which raises your cortisol levels to the roof; your inbox piles up; the train is full—and late; and you're having problems with your boss. Maybe this is our lives in a nutshell.

Stress means that the adrenal glands work all the time and the body is flooded with sugar, while the ability to absorb this sugar with the help of insulin is diminished. Meanwhile, inflammation, in turn, drives more cortisol, which continues to drive inflammation.

This body needs to breathe, with body and soul!

That's why anything we can do that counteracts this cortisol boost is not just a pleasant moment for the soul to catch its breath. It's more healing than that—it's a way to cancel the risk of inflammation.

New studies, at Carnegie Mellon University, among others, show that as little as a three-day course of meditation and mindfulness lowers the level of the inflammation marker IL-6. This is in contrast to common relaxation exercises, which didn't have nearly the same positive effect on the participants (all of whom were stressed-out professionals in midlife).

The practice of meditation and mindfulness is more than simple relaxation. It's a more focused process that has a deeper effect. Similarly, deep-breathing exercises have been shown to decrease inflammation—and this is where yoga comes in.

Studies of both healthy people and breast cancer survivors show it clearly: Long-term yoga practice is a supercure for inflammation that brings down the levels of pro-inflammatory cytokines like IL-6, TNF-α, and IL-1B. For breast cancer survivors, it was also clear that vitality increased among the subjects in the yoga group.

The woman at the bliss workshop and I talk about this, among other things. We talk for a long time. She will now try to find her way, and she decides to begin by looking through the assortment of meditation apps. It's a good thing that there are so many new tools available now.

But the greatest meditation of all doesn't require an app and is as old as life itself. Or as the Dalai Lama himself is supposed to have said: "The best meditation is sleep."

Many years ago, when I was a science reporter at a Swedish television station, I worked on a documentary with a sleep professor at Karolinska Institutet, Torbjörn Åkerstedt, who was investigating stress and sleep.

"Remember," he said, "that sleep is the most complete recuperation process that a body can engage in, packaged into one single little package."

We all know how we feel the morning after a bad night's sleep—vulnerable and uncomfortable in our bodies. Does sleep have a restorative effect on early inflammation as well?

Research published in *Best Practice & Research: Clinical Endocrinology and Metabolism* by the neurologist Janet Mullington and others, shows, for example, that a lack of sleep leads to increased deployment of cyto-kines, or pro-inflammatory markers. It seems that sleep is yet another piece of the puzzle.

I'm thinking about my conversation with Torbjörn Åkerstedt. He said one other thing, "Try to get to sleep before eleven o'clock. Take advantage of the natural cortisol dip that occurs then."

The body experiences a natural decrease of cortisol at around eleven o'clock at night, a rhythm that was created for another world, before elec-tric lights and the possibility of binge watching Danish mystery series on TV. This was a world where people lived by natural light and fell into deep sleep at eleven o'clock in order to wake up with the animals at dawn.

"If you can pinpoint the natural cortisol dip, you can get a greater re-storative effect from the same amount of sleep," said Professor Åkerstedt.

Which makes me think of an Ayurveda doctor I met in Kerala.

"You must please go to bed, please madam, please at ten at night," he said. Then he shook his head, like our "no," as many Indians do when they mean "Yes, this is important."

I begin to see the pattern: when I go to bed earlier, I have a better day. And this is probably connected to a generally lower inflammation level in my body. The trick is to be more disciplined about going to bed. You have to shut down all the beeping, pinging screens, stay away from Instagram and bouncy Twitter feeds after nine o'clock, and just take it easy. That's hard sometimes. And you can't always just think about discipline.

I'm still in Los Angeles when another thing comes to mind.

One spring I have a terrible cold, the kind that seems never to go away, a tough virus that settles like an iron claw in my throat, with a constant, low-grade feverish sweat. Meanwhile, I have a lot to do: way too much work that has to be done before midsummer, high pressure to deliver, a problem within the family that worries me, and a friend going through a hard time. I feel anxious, pressured, tired, and ragged.

With my new frame of reference, I can see my cortisol going through the roof for all kinds of reasons, my cytokines hovering in the background.

I feel completely wrung out and decide that I can't continue like this. There's too much bickering; too much energy and willpower are being consumed by having to constantly plan for my new lifestyle, all the extra effort. I feel nervous and down, like a mangy cat after a fight. I simply can't make any more effort or deal with any more obligations beyond the minimum needed to take care of my job and family.

When I gradually get wind in my sails again, I decide to go to church and find some awe. My eye lands on a passage from the Ecclesiastes.

"To every thing there is a season, and a time to every purpose under the heaven: A time to be born, and a time to die; a time to plant, and a time to pluck up that which is planted; A time to kill, and a time to heal; a time to break down, and a time to build up; A time to weep, and a time to laugh; a time to mourn, and a time to dance."

The ebb and flow of life.

It strikes me that the idea of peace and stillness also contains an acceptance of the shifts and rhythms of life, of the reality that sometimes we are old and frail, and that my body, like everyone else's, has its limits—even if we do often have more energy than we think. We often get uncreative and think that resting has to mean pulling a blanket over our heads and lying completely still.

I realize that an anti-inflammatory lifestyle also has to include a large measure of humility in the face of life's "disease lottery," and that this concept must never be used to hold a judgmental view of human beings in which strength and health somehow are morally better than frailty and illness.

That's why I don't like to describe food as clean or unclean; it can all too easily become a metaphor for the person who eats the clean or unclean food, so that, for example, "clean" people eat salad, fish, and probiotics—no, I don't agree with that. Nor do I like the concept of cleansing, or detoxing, since it implies that food is somehow unclean and that a person who eats "unclean" food is also a bad person.

Such a view of human beings will only create a new type of classification or lead to blaming people who are struggling with health challenges, or people who, for various reasons—powerlessness, lack of energy, poverty, vulnerability, illness, genes, upbringing, cultural tradition, preferences, or other reasons—eat in a different way than I do, regardless of whether they choose this lifestyle, feel forced into it, or just think in a different way.

My vision is instead an idea that all of us have a responsibility to take care. We have received a body from life, our parents, God, or the great energies, and it is ours to use as best we can. That's why lifestyle should never be a goal in itself but should instead be used to create value for the larger whole. I don't hold any judgments about how other people live, but I do feel a joy in sharing my new insights. Or to be completely honest—it's so damn exciting that nothing can stop me. I just have to share them!

This insight also includes the realization that sometimes the body shouldn't be used at all but should just rest.

One of my friends has a mother named Irma, a stylish and warm grand dame who has always been a very active and hardworking woman. Sometimes, when my friend was growing up, Irma would indulge in a day when she had the so-called "Irma illness." This meant that she just retreated and lay in bed with the blanket pulled over her head. She stepped away from the world for a day. And maybe we all need our "Irma days."

But we also need our "Irma days" filled with activity. Simply put, we need our flow, our rhythm, as the Prophet says. A time to be born, a time to die, a time to be active, and a time to be still.

And now it's time to fulfill my promise to myself. Carrying old, negative feelings for too long isn't respectful toward myself, nor is it anti-inflammatory. I want to own my new knowledge and my ambition to leave this burden behind me before I return home from the United States. But who will help me along the way?

I go to a gym in Los Angeles wearing my newly purchased Bliss sweatshirt. And suddenly she's there.

"Were you at the workshop too?" she asks.

"Yes, you too?"

Her name is Arriane, and she's a life coach who helps people with both private and professional issues. I immediately get the feeling that she was sent to help me. We decide that she'll come and see me at the place where I'm staying so that we can talk about what to do.

She comes on a Thursday afternoon, and after I've told her what's bothering me she says:

"We can get rid of this. Do you want to?"

Without going into any personal details, I'll just say that this empathetic and wise coach takes me through a number of steps that might seem like therapy preschool for a person more used to therapy. For me, it's a pioneer landscape; I ride in my wagon through hitherto unexplored territory. After two hours, I feel dizzy and exhausted, but my mind has been cleared.

She leaves, and I sit there thinking about how I can complete this process of letting go.

Eventually I make my way down to the Pacific Ocean and the beach by Santa Monica. It's a wonderful evening, with golden, glittering sunlight dancing over the ocean's rolling waves. I sit down on the beach and write a letter about things that I need to let go of.

The idea comes to me that I want to capture this moment of giving up my burden to the sea, so that I'll never forget that this is where I left those sad, gray feelings of devaluation, shame, and guilt. I look around. To my left there are some Frisbee players, and to my right, a couple in love. I don't want to disturb them. But farther away, from the beach by the pier, a jogger is approaching.

She's going to stop right here, I think. Then I push away the thought. She's running at full speed. Why should she stop right here at my feet?

She stops, right at my feet.

"Hi," she says, smiling.

"I have a letter that I'd like to give the ocean. Can you take a picture of me at the same time?"

I lay the note in the sea. It bobs away like a little *Kon-Tiki* raft laden with heavy sorrow that must now live its own life without me.

Goodbye!

She takes several pictures and tells me how to stand so that everything will fit in the frame.

When I go up to her, she looks me deep in the eyes and gives me a hug.

"Just let it go," she says, smiling with her blue eyes.

I tell this remarkable story to my actress daughter, who after two years in Los Angeles is open to all the new spirituality. She looks at me quietly with her big, deep brown eyes.

"Now you're ready to move on," she says. "Be strong, live your truth, Mom."

I promise her I'll do just that.

*Suddenly all my ancestors
are behind me. Be still, they say.
Watch and listen. You are the
result of the love of thousands.*

—Linda Hogan

13. ROOTS

Now it's time to finally dig down into the mystery of why all of this works and to fit together the remaining pieces of the puzzle so that we can see the whole picture. We have to look further back into the roots—my roots, your roots, all of our roots.

If I could have lived a few other lives, I would have spent at least one of them as a very adventurous paleoanthropologist, a cross between Indiana Jones, Karen Blixen, and a glamorous Egyptologist I saw a long time ago in the not very classy film *Sphinx*. (Her name was Erica, which incidentally inspired the name of our oldest child.) In this life, I would have traveled to East Africa, to the Rift Valley, to try to uncover the ancient bones that rest there as a witness to human mysteries.

Deep inside the earth's crust, there is constant movement as the tectonic plates shift. That's why the continents today don't look like they did 225 million years ago, when just about all of the earth's existing mass was pressed into one giant continent in a large, desolate sea. As the tectonic plates collide and pull away from each other, mountain chains and rifts arise in the edges between them. In East Africa there's one such wound in the earth, a crack that can even be seen from space. It's called the Rift Valley, and it runs through Kenya's dizzying western mountain country, past the border of Tanzania and Uganda, up toward Somalia and Ethiopia. Here lie majestic old volcanoes, like Africa's highest mountain, Kilimanjaro, with its snow-capped white peak.

A long time ago, the volcanoes were active, spewing out masses of lava and ash. This produced layers in the earth, allowing scientists to use radioactivity to measure precisely how old everything is, including human bones. There are bones here that are millions of years old and can tell the story of human beings and their journey from humanoid ape to the beings we have become. This has always fascinated me, and it was the reason I wanted to study genetics at one point.

We are in Ethiopia—more about the why later. But right now, I'm standing in Ethiopia's National Museum, in front of a very small person, a being about three and a half feet tall. Her name is Lucy.

I walk around her, examining her. She's both ape and human, standing upright on two legs like a human being but with long arms like an ape. She has a lower forehead than we do, but stronger jaws and teeth.

Next to her lies the skeleton that the reconstruction is based on. This skeleton is all of 3.2 million years old. It was put together from hundreds of little fragments of fossilized bone that were found one Sunday morning in November 1974 by the paleoanthropologist Donald Johanson, near the village of Hadar, in the Awash valley of the Afar Triangle in Ethiopia, in Rift Valley.

Johanson wrote a book about how he found Lucy, and I still remember my excitement when I sat in my apartment in Lund and read it. (There are unconfirmed rumors that the anthropologists got high on LSD after the experience, and that Lucy was named for the Beatles song "Lucy in the Sky with Diamonds," which is considered to be a code for LSD.)

The archaeologists realized right away that the discovery was sensational, since they had found a skeleton that was (a) unusually complete, and (b) the oldest skeleton of a hominid that had ever been found.

She was named Lucy, as well as Dinknesh, which means "you are wonderful" in the Ethiopian language Amharic. She was also given the scientific reference name AL288-I. A beloved child has many names, as the saying goes.

So why are we in Ethiopia?

I'm here because I have a friend with connections to the country and because this friend once, when I was homeless in New York during my student years, let me live on his couch, for a whole month, right in the middle of Manhattan, without asking me to pay a penny. Now this friend is going to get married here, to his stunning Finnish girlfriend, an expert on antiquities. Some of us loyal friends have flown here to attend their wedding.

On the long plane trip, we make a stop in Jeddah in Saudi Arabia, in order to admit an inspector who's going to check whether the women who are getting off the plane are covered enough and whether any of the passengers have alcohol or pornography hidden in their seats. I don't like his rigid face, his imperious manner, even less what he represents, and I shout:

"My seat is full of everything forbidden, take me if you can!"

No, unfortunately I'm much too well mannered to do that. I'm also all too well mannered when we arrive in Ethiopia and meet a representative of the rock-hard dictatorship that's keeping the opposition in an iron grip, and all I do is quietly wonder why the country doesn't allow free elections.

"You Europeans always have to nag about democracy," he says harshly.

Yes, we like to do that.

But these unique and generous days also give us memories for life. The Coptic priest turns up at the last minute with his retinue of black-clad advisors, a court full of drama, as if lifted out of a Shakespeare play. The bridal couple dances a traditional Ethiopian wedding dance. We visit the grave of Carl Gustaf von Rosen, the pilot who dropped food bombs on the starving Biafra from the air, which I still remember from my earliest childhood. And we joke and laugh and talk like friends do.

Here we are then, in Addis Ababa. A cornflower-blue sky and crystal clear, shimmering air reflect its location in the highlands, while the hardworking bustle on the streets reflects its great poverty.

But all of this becomes a side narrative when I'm standing in front of Lucy in Ethiopia's National Museum.

Lucy is a representative of the long ancestral lines of humanity.

She isn't a *Homo sapiens*, like we are, because we had not yet begun to exist when Lucy was alive. She is an *Australopithecus afarensis*. She's said to be twelve years old, which in Lucy's day meant she was an adult. You can see that from her teeth, which are fully grown. Her hands are free. I walk around this being, examine her through my anti-inflammatory glasses and wonder how she lived, because I think she holds the partial answer to our question.

What are your days like, Lucy?

Who are you? What do you do? What do you eat?

Perhaps she might answer like this:

"It's a hard life in many ways, filled with hunger and danger on the savanna. We live in the borderlands by the forest and take what we find; today it was just turtles and crocodile eggs. We eat nuts, seeds, green plants, and sometimes meat. Sometimes there isn't enough to eat."

She might add this:

"I move around for a few hours a day while I'm gathering food but often relax in the warm sun. I sleep when it gets dark and wake up at dawn."

There are periods of rest and periods of feverish activity. She has natural fasting periods when there's less food. She doesn't eat any refined carbohydrates but does eat organic proteins, good fats, and lots of polyphenols and soluble viscous fibers, a diet that my researcher contacts would have approved of.

In a way, she lives an anti-inflammatory lifestyle, with very little gluten and lactose, and plenty of good bacteria that she ingests through her food. She lives with natural exercise, a life in nature, in close contact with the magical night, where the starry sky is so clear that you can almost touch it. She lives in the presence of Africa's rosy sunrises and vast open spaces. She lives in her little flock, completely dependent on cooperation with others to survive.

Is Lucy's existence the key? Have human beings developed and their genes been refined to suit this "Lucy life"? And is that why we develop inflammation in our modern lives, in today's society that is so different?

But human beings, of course, did not stay on the African continent. We began to migrate.

Hundreds of thousands of years ago, a group of early humans walks away from what is today the African continent. These people evolve to become Neanderthals, among others.

About two hundred thousand years ago, Lucy's descendants become Homo sapiens, sapient beings. It doesn't happen overnight, but our fore-mothers and forefathers begin making tools, tame fire, and learn to speak a more sophisticated and symbolic language than the guttural sounds they are thought to have used earlier. They spread out toward Central Asia, where some of them go west, to what will become Europe. Others go east, to what will become Asia, and on to America. They take with them both our genes and our methods of survival.

Someone gives me this image as an illustration—maybe it's Rita, maybe someone else. *How does the flock eat along the way?* Do you have McDonald's, or hot dog stands? No, there's nothing but savanna, steppes, sea, tundra, forests, impenetrable mountain areas, flowing rivers, and deep, stony ravines. People are completely dependent on their ingenuity, their survival instincts, their strategic thinking, and their planning for everything they need, including food.

They hunt antelope, fish using simple spears, gather walnuts and roots, and they also keep a tight grip on their ever-present hunger so as not to eat everything at once. Instead they discipline themselves to dry meat, store roots, dry berries, and gather the nuts they've found along the way, perhaps in little bags made of animal skins. This is strategic eating, and it gives sensational results, since Lucy's descendants will colonize this entire magical, blue-green little planet and go from being one of many animals to eventually assuming the throne of evolution.

I stand before Lucy, thinking about these ancient mothers and fathers. What would Lucy think of our eating habits? What would be the greatest difference between our diets—sugar?

On the savanna, there's almost nothing sweet, except that Lucy might have come across an occasional beehive with honey on a lucky day. Lucy is built for nutrient deficiency and sugar deficiency. But because sugar is dense in nutrients, if she finds it she will want to eat as much of it as possible and as quickly as possible, without stopping. After all, the next beehive might never turn up.

But imagine that you put the little furry Lucy, her genes shaped by the message to constantly look for sweet things, in a shiny grocery store where the shelves and counters are overflowing with jam, cookies, and ice cream. You only have to reach out and take it (and then pay). She would probably go just as crazy as a five-year-old does over the boxes of candy in that store. This is what many of us grown-ups do as well, only with a bit more social finesse. So we eat too much sugar because the Lucy within us is constantly craving what was so scarce on the savanna.

The thought occurs to me that in a way, I *am* Lucy. Well, not literally her being, this person who died much too young from what scientists think was a dramatic fall from a tall tree that broke her shoulder and pelvis. But I, or we, share her basic biochemistry. Lucy is like a shadow image in our own genes.

This is where we must go to continue digging. And that journey will now take a step up to a more hard-core scientific level, for the reader who is so inclined.

When I was studying at Lund in the 1980s, at the genetics department where Albert Levan discovered that human beings had forty-six chromosomes, people saw genes as the ultimate recipes for everything from eye color and height to digestion, heart rhythm, and, to some extent, temperament and behavior as well. These recipes together made up the cookbook of life and were in principle impossible to change, unless big mutations occurred.

Just as a regular cookbook consists of endless variations on a number of basic ingredients, like meat, butter, spices, vegetables, and so on, that are combined again and again to make meatballs, hamburgers, or meat stew, genetic recipes are a constant variation on just four raw materials: nucleic acids.

The ingredients A, T, C, and G are scientists' abbreviations for adenine, thymine, cytosine, and guanine. They are attached along the DNA chain, and the recipe is read so that the substances are decoded in groups of three. When the recipes are read you will have, for example, AAA, ATC, GTC, CGA, and so on.

Each unique triplet codes for a unique amino acid, of which there are twenty-four. The different amino acids are linked to each other like hundreds of beads along a magnificently long necklace to become a complete protein.

TGG is the recipe for the amino acid tryptophan. AAA codes for lysine. So if a long gene sequence begins with TGG and then has three nucleic acids AAA in a row (which is lysine), in the end a long protein will form that begins with tryptophan lysine and then continues to be built in a long chain, according to the recipe. This translation, from the four letters to amino acids that become long proteins, is the core of life. The proteins produced are often called the workhorses of the cells because they do all the work.

Anyone who's given birth knows that newborn babies are pricked after birth in order to obtain a blood sample. Among other things, the sample is checked for a specific gene, the gene that produces a crucial enzyme and sits on the long arm of chromosome 12. Without this enzyme, a newborn baby runs the risk of developing the serious hereditary disease phenylketonuria, or PKU.

Every action, every thought, every feeling, is founded on this process of protein formation working. The body doesn't understand any other language but this—let us call it the biochemical language. Genes are involved in everything. This can be hard to absorb for someone who feels that human beings are mainly shaped by their environment, but that's because the process is often misunderstood.

The fact that everything must be translated into biochemistry doesn't mean that your DNA ultimately decides everything. On the contrary, there are many things that affect genes from the outside—both negative things, like threat and stress, and positive things, like security and kindness. Genes are constantly interacting with the environment.

Even if threat, stress, security, and kindness come from outside, all such outer events must be sensed via the body's signaling system and then interpreted and turned into some kind of response. In this process, genes are the main players. A kind of highly effective reader (which we can call mRNA) and another professional cook that we'll call tRNA constantly walk along the strand of DNA, reading recipes and making new proteins by mixing amino acids that are then placed one after the other to form long chains.

When I was studying genetics, we learned that this was the whole story, that this cookbook chugged along steadily, and that the recipes were read and prepared like a kind of mechanical process. That's what we students learned. But behind closed doors, scientists were perplexed.

The question they were thinking about was this: How could you reconcile the idea of stable genes on one hand with the ever-changing human being on the other?

Think of your own existence. From newly fertilized cell, through the entire development in utero, via birth, to human life as an infant, child, growing teenager, adult, and aging person. This is you; your genes are the same all the way through.

Yet your human form constantly takes new shapes. This is what scientists didn't understand. What was happening? If those genes really were so stable, how could they use the same recipes to produce beings who behaved so differently during different phases of life? And how could the DNA in an eye cell look exactly like the DNA that was deep inside uterine cells? Why could one cell detect the shifting colors of the summer sea, while the other one was a mucous membrane to which an embryo was attached? The same recipe yielded such different results.

On top of this, there was the great battle of nature vs. nurture that raged all through the twentieth century, between those who considered biological factors to be all-important and those who saw environmental factors as the key.

Eventually a more sophisticated view of this conflict emerged in scientific circles. One thing that contributed was the large twin studies that were done where identical and non-identical twins were compared to each other. Twins are ideal objects of study, because identical twins have exactly the same genetic material. If they were separated at a young age, you can compare how their lives, illnesses, and behaviors fluctuate compared to identical twins who grew up together. This gives scientists a good way to measure the respective roles of heredity and environment.

It turned out that certain traits were completely controlled by genetics, like eye color. Other characteristics needed an interaction between biological and environmental factors in order to develop. However, the body only speaks biochemical language, so even those things that are shaped by the environment require an interaction with the genome. There had to be a window that could be opened between DNA and the world around it in order to let them interact.

There had to be something more than just genes.

That's why I'm standing outside the Karolinska University Hospital in Solna, Sweden, one day in May. There's a strong wind and it's snowing—in May!

Some researchers are standing in their shirtsleeves in front of the Center for Molecular Medicine, smoking. They are muttering in English that Sweden is unbelievably cold, and how can people stand to live here all their lives? "What a bloody country."

I'm on my way to meet Tomas Ekström, a professor at Karolinska Institutet. He's working in the new, fast-growing field of study known as epigenetics, which is trying to find answers to the great genetic questions. This work is the front line of research, precisely the front line that our lecturer at Grayshott was trying to teach the gut group about, although that was at the level of an elevator speech. I'm curious about Professor Ekström and his epigenetic front line, and this time I want more information than what can be transmitted during an elevator ride.

Professor Ekström greets me; he has a steady blue gaze and is wearing a shirt and jeans, a cup of coffee in his hand. We sit down in the room next to the break room, where large, heavy literary works are on the shelves. James Joyce, Ayn Rand . . . apparently his group of researchers' interest in the existential is not limited to test tubes.

The whiteboard in the room looks like a parody of science lingo. There are long chemical formulas, diagrams of how molecules are transformed or broken down, and questions with five question marks at the end.

Professor Ekström dives right in, and he challenges the questions that I sent him before the interview.

"You've posed questions that are informed, but you've posed them the wrong way," he says.

"How so?" I wonder.

"I'll have to give you some more background," he replies.

He thinks for a moment.

"First, you have to understand that epigenetics is about qualities in the genetic mass that are inherited from cell to cell during cell division but do not affect the gene sequence itself, but rather the way it is expressed," he says.

"I think you'll have to find a better definition," I say. "Why is epigenetics important?"

"Okay, I 'll try again," he says. "How about this: epigenetics is the window between the inherited genes and the environment."

That is why I'm here, in order to understand what the genetic window to the outside world looks like, and how genes can be affected by lifestyle. This science takes its name from the Greek word *epi*, which means "above" or "next to."

"This is central," says the professor. "It means that there's another control system that's superimposed on the genes and affects how they're used."

"But how exactly does that work?" I ask.

"Come with me," he says.

We go out into the hallway, where Professor Ekström shows me a large poster on the wall. In the middle of the poster is something I can only describe as an octopus. There's a large blob in the middle, with thin arms that stretch outward.

"Look here, in the middle you have the histones," he says.

Histones . . . I root around in my memory to see what I can remember about them as I look at the octopus. Where did they fit in again?

All of us contain huge amounts of DNA. In a single cell, there's enough DNA that you could stretch it out two or three yards if you were to combine it in the form of a clothesline. If you fold together all your DNA to form a single, long thread and throw that thread up in the air, it would be long enough to go back and forth between our planet and the sun three hundred times. It's staggering, and the DNA needs to have an incredible ability to change shape, roll up, or dissolve.

Sometimes it's tightly wound, like sewing thread wound tightly around a spool, or like a stiff bun worn by a churchgoing lady in the 1800s. Sometimes it's long and tangled, like someone who's been sleeping on sweaty sheets and wakes up like a fuzzy troll. It's mainly during the process of cell division that DNA is reshaped—from tight knot to fuzz, and back again to a knot. The central point that the knot is tied around is eight histones, the center of the knot. The DNA strand is coiled exactly 1.65 times around this center. It is this group of eight histones, called nucleosomes, that form the head of my octopus when I look at the professor's poster.

"And the arms?" I ask.

"They are the histones' arms. That's where it happens."

He points to the long octopus arms, and I can see something like little control systems. An amino acid here and another one there, like keys on a keyboard.

Professor Ekström shows me how nature attaches small groups of chemicals onto these histone tails. Apparently, this is called methylation, phosphorylation, acetylation.

"Meaning what?"

"This is one foundation of how the environment has an effect," says Dr. Ekström.

"How so?"

"When the tails are methylated, for example, the tightness of the DNA coil is affected. It can become more loosely or tightly packed."

Here we have the methylations, the ones that Dr. Fraser in Loma Linda was tracking in his research.

When the DNA becomes more tightly or loosely packed, the readers (mRNA) and the cooks (tRNA) are affected. When the DNA strands become longer and curlier, these busy helpers have easier access to the recipe book, which earlier lay tightly coiled like in a vise. Likewise, if little groups attach to the histone arms that make the DNA coil more tightly around the histones, the mRNA or tRNA can't access it as well, and the recipes become different.

"This is the process that determines if genes that you carry will be expressed or not."

Let's stop and think about what this means.

There's a system on top of the genes, a kind of master control, or conductor, that determines whether the DNA is going to be tightly wound like the hair of a nineteenth-century lady or in a freer, fuzzier style and thus decides if the genes are quiet or active.

And the object, of course, is to activate the genes that are responsible for building up the body and to make the genes that contribute to illness less active.

But how does that happen?

It's been known for a long time in the scientific world that organisms that get too little food, that starve or partly starve, live longer. Starvation can be good.

The question is why.

When two researchers, Leonard Guarente from MIT and David Sinclair from Harvard, studied half-starving yeast cells in 2003, they found something that many people considered sensational. They found a group of master regulatory genes, the so-called sirtuins, that functioned sort of like an old-fashioned telephone exchange. When there were too few nutrients around the little floaty yeasts, the exchange lit up and the sirtuins were turned on.

There are seven sirtuins, and all mammals have them. Sirtuins are epigenetically active; that is, they regulate which other genes will be activated or deactivated, in the way that I just described.

One hypothesis was that the sirtuins can react to mild external stress, pushing the organism to do more and better. The phenomenon is called adaptive stress and is sort of like when we have a deadline at work. We become faster-thinking, focus on delivering, and are less dependent on food and sleep. This aspect of short-term stress makes us perform better. (Of course, it's quite a different thing if the stress continues for too long; then it breaks us down instead.)

What scientists suspected was that active sirtuins not only can extend life but also have a rejuvenating effect, in the sense that they increase the chance of a long life by changing the metabolism, increasing muscle building, increasing cellular repair mechanisms, and so on.

Could this be an explanation for why people who fast regularly, for example with the 5:2 method, not only lose weight but also experience greater well-being in general?

But then came the bomb, in the science world's most-respected journal, *Nature*. An article there questioned whether the sirtuins really had the powerful effect that had been claimed. Researchers around the world had trouble replicating Sinclair's and Guarente's results. Now eyebrows were raised and the research disputed, since reproducibility is an absolute requirement for something to be accepted as true. (Other researchers should be able to produce the same results if they set up an experiment in the same way.)

Were the sirtuins a false trail?

Then, in 2016, there was a new study, which in itself was a sign of the times since it had been carried out in China. Scientists there, directed by Nannan Zhang from the elite Tsinghua University in Beijing, had achieved interesting results. (Only twenty years ago, these Chinese researchers had not had enough resources to compete on the international scientific front, but now, with China's emergence as an economic superpower, they suddenly have the opportunity to participate in breaking knowledge barriers.)

In their study, they continued to build on the old findings, that is, that starvation increased longevity among the species studied, both yeast bacteria and mammals. The Chinese scientists pointed to a special sirtuin or master regulator—SIRT6—that was responsible for transferring the effects of starvation, in part by affecting a factor called NF-kB. This factor—and now we're really getting hot—is active both in inflammation processes and aging.

This team now demonstrated that if you decrease calorie intake in mice, the capacity of their kidneys as well as the power of SIRT6 increased after six months of treatment. They also showed that cell cultures with a low sugar content had better SIRT6 expression, and lived longer, than cell cultures that received more sugar. The team could also see that the more SIRT6 "talked," the more aging was delayed in the organisms studied. And conversely, when SIRT6 grew quiet, the cell cultures aged at a rapid pace.

The sirtuins were back in the spotlight. Here, finally, was an epigenetic explanation for the positive effects of food deprivation on longevity and decreased inflammation—an explanation of the effects of the 5:2 diet and its anti-inflammatory influence.

But for the vast majority of people, life doesn't work if you're constantly going around half-starved. What can you do, then, if you want to imitate the effects of food deprivation without having to feel hungry?

One of the hypotheses that are now circulating about polyphenols is that they somehow "imitate" the effects of food deprivation precisely by affecting the sirtuins and that anti-inflammatory foods contain substances that in turn affect the sirtuins in a similar way that food deprivation does. This is without having to starve; people can eat until they feel satisfied.

Then perhaps the polyphenols in our rainbow of berries and vegetables might actually affect the sirtuins. For example, people have found proof that resveratrol, which is found in red wine, for example, affects another gene in the master regulating SIRT family, namely SIRT1.

It's also been shown that mice who get resveratrol (or "drink wine") while eating a super-fat diet manage to live as long as mice who live on a more spartan health diet but don't get resveratrol. Something in the resveratrol thus seems to be able to even out, or even compensate for, the defects in the mice's diet.

And yet . . .

There are many different kinds of anti-inflammatory foods. It's possible that the theory about sirtuins can be linked to the rainbow and polyphenols. But that doesn't explain the effects of omega-3 fats, soluble fibers, or probiotics. I still haven't met anyone who can say, "Voilà. Here is the whole picture, this is how it works. We're all done." It simply has to be a matter of many different mechanisms working together. Doesn't it?

It's like looking at a big painting through little holes in the wall. I can see pieces of the image, but no one has yet been able to see the entire painting. The arguments are built up like in a tricky criminal investigation, where there were no witnesses to the murder, yet the corpse lies there. You suspect certain things and have to compile evidence that points in a certain direction, while understanding that there are still more leads to be found.

I make a large diagram on a whiteboard. Just like our detective, I stand before my own evidence board late one evening. I've put up a number of pictures and Post-its.

On my Post-it notes:

- The connections (strong) between inflammation and disease
- Lifestyle and longevity in the Blue Zones
- Research at Lund about anti-inflammatory food
- Yoga practitioners look younger
- Research on longevity and semi-starvation, the sirtuins' epigenetic effects (email Sinclair!)

Also: A picture of a wine bottle, a picture of a bottle of probiotics, and a happy salmon.

All these ideas and bits of evidence are coming and going through my brain, like ants in an anthill. How will I be able to make sense of it all?

"It's been known for a long time in the scientific world that organisms that get too little food, that starve or partly starve, live longer. Starvation can be good. The question is why."

Inside you one vault after another opens endlessly. You'll never be complete, and that's as it should be.

—Tomas Tranströmer,
For the Living and the Dead

14. ONE VAULT
AFTER ANOTHER

Meanwhile, other things are happening in my life.

There's my husband. He's made the entire journey from mildly uninterested, to mildly opposed, to active challenger of this whole new lifestyle. Like all of us human beings, he is capable of great flexibility and is usually wonderfully warm and accepting, but he can also be quite stubborn when he doesn't like something. All the physical manifestations of my new lifestyle are firmly in his "don't-like" zone: the big bags of protein powder and nuts, the anti-inflammatory pills, jars of seeds, omega-3 and BCAA supplements.

"Okay, I admit, it's a lot of new stuff. But it's my food," I try to explain.

"Look at that little shelf down there," he says. "There's my food."

By "my food" he means all the things that are needed for a good, traditional breakfast with toast and eggs. Coming from the region of Skåne as he does, he feels at home in the deli area with its traditional gold mine of sausages, liverwurst, and cheese.

But he also likes running. Now he's going to run a half marathon in the London parks, between Green Park and Hyde Park, on Saturday morning. On Thursday he comes to me.

"What kind of food should I eat that will give me good energy?" he wonders carefully.

Instead of his loading up on pasta the night before the half marathon, I make him salmon grilled with turmeric, sweet potatoes with rosemary, roasted vegetables, and an arugula salad. For breakfast he has a large smoothie with protein powder, almond milk, blueberries, spinach, nuts, flaxseeds, and two big spoonfuls of raw coconut oil, omega-3 pills, and probiotics. He runs like the Roman god Hermes with wings on his heels. He bounces along in Hyde Park and glides across the finish line more easily than runners half his age, in record time.

"That went incredibly well," he says, surprised. "I felt so light in my body." And that's where it begins.

He changes his breakfast completely and wants to have a smoothie every morning. Soon he begins making it himself. One of our sons psychs him up to work out at the gym. After a month he looks more energetic than he has in many years, and he gets compliments on his new, younger appearance. We call the smoothie his "looking-good smoothie," and my husband goes from being anti-powder to actively hunting for protein powders on the shelves. The interesting thing is that this turnaround happens after I've already given up and gone from being a preacher to being a calm helper.

At the same time, more and more friends around me begin asking about my new lifestyle. I often get phone calls, and I send out food lists and recipes. They test it out and see that it works for all kinds of aches and pains. Everything from one friend's panic attacks to another friend's joint problems—they both get better with an anti-inflammatory lifestyle. And I have contact every week with my dear friend whose illness has made her rethink her lifestyle and diet. We exchange thoughts and articles and test things together to see what might help her. I see all this and do a lot of thinking.

And how are things going for me personally? Well, I'm keeping up the lifestyle, but I'm less picky and follow my instincts more regarding what and when.

If I feel swollen and slow, I try a mini-fast for sixteen hours and do some light and quick exercise instead of lifting weights. If I feel anxious and frozen, I increase my food and lift heavy weights to ground myself. Too much aggression and stress? Maybe I need to meditate, practice yoga, go vegetarian for a day, or listen to my spirituality app.

Above all: I worry less about what other people think of my lifestyle. Instead, I'm more interested in my own path and how it's linked to the greater whole, the greater good.

I learn not to judge my own eating as good or bad but just accept that some days I might drink too much wine or eat some birthday cake at a friend's place without "ruining" that day. I accept the situation for what it is and don't keep putting myself down but calmly go back to an anti-inflammatory lifestyle and the healing that it brings.

Then, I send off an email.

The answer comes after a few days, from Harvard University in Massachusetts. Professor David Sinclair, who once discovered the sirtuins, has replied.

"Does the evidence still hold up?"

According to Professor Sinclair, the model is correct, and they still believe that the activated sirtuins affect the histones and thereby also the genes. I'm curious to know how far he wants to take the conclusions of his study.

"Can we replace medicine with anti-inflammatory food?"

"The anti-inflammatory approach is interesting and works to prevent illness, but it has a long-term effect on the body and might not be enough once you're already ill," writes Dr. Sinclair.

He believes that medicine cannot be replaced with food in the case of a serious illness.

"If you have diabetes, you need to take medication. Working with anti-inflammatory food takes decades."

That sounds reasonable.

I look for other researchers who might provide an additional evaluation. My old contact Barry Sears in the United States launched a diet thirty years ago that in some ways resembles the anti-inflammatory diet, the so-called "Zone Diet." His books about it were on the *New York Times* bestseller list for months.

I reach him on the phone.

"The biggest difference between us and our ancestors," he says, "is that our balance between omega-6 and omega-3 is completely wrong."

The omega fats are usually called essential fats. The body can't produce them by itself, so they need to be added from the outside via food. They complement each other, and both are needed for the brain, heart, and joints. Omega-6 exists naturally in certain seeds and nuts, as well as in hydrogenated vegetable fats, like in industrially baked cookies and processed foods. Omega-3 is often found in fish that feeds on krill, and in some nuts and seeds.

In our original environment, during Lucy's time, the balance between omega-6 and omega-3 in our diet was about 2:1, with twice as much omega-6 as omega-3. Today the ratio might be as high as 20:1, or twenty times as much omega-6 as omega-3.

"If we adjust that balance, we can adjust much of the inflammation," says Dr. Sears.

Omega-6 has a building function but can also be broken down into inflammation-driving substances. There are two types of Omega-3, EPA and DHA, which counteract inflammation. But how?

I look into it and discover two ways.

For one thing, omega-3 counteracts a destructive product from omega-6 fats, the so-called eicosanoids. Omega-3 fats can also form substances that directly or indirectly counteract the cytokines and, in particular, the very potent inflammation driver NF-kB, the cunning little agent we encountered earlier in connection with the sirtuins.

NF-kB swims around in our cells, and when there's an external stressor, it moves toward the cell nucleus at lightning speed, encouraging the genes to produce more inflammation-driving substances. You don't want to get on the wrong side of NF-kB, which is a master regulator when it comes to stirring up inflammatory genes.

And bingo. Here we find sugar, which triggers NF-kB activity and thus also the inflammation-driving effect of the genes.

And bingo again. Through a Twitter source, I find a large summary in the journal *Frontiers in Immunology*, where the authors have compiled studies from the past eleven years about meditation, yoga, and tai chi, or what researchers here call MBI (Mind Body Intervention), and how such exercises affect the genes.

The overview shows that someone who regularly practices conscious relaxation decreases their production of NF-kB and thereby also the accompanying inflammatory gene expressions.

GOOD FATS

Fat is your friend, since it provides satiety and taste and controls the GI value of other food. Omega-3-rich fats are actively anti-inflammatory, but you should keep omega-6 fats to a minimum, as they are building blocks of inflammation.

- Avocado
- Butter—organic, from grass-fed cows, which contains plenty of omega-3
- Butter/oil spreads—butter and rapeseed oil, or butter and olive oil are good combinations
- Coconut oil—virgin or raw
- Fatty fish—contain extra amounts of the long-chain version of omega-3, the one called DHA and EPA, and are inflammation healing to the max
- Nuts and seeds
- Olive oil—as dark green and raw, or virgin, as possible
- Olives

Let's repeat that: Yoga and meditation change the gene expression and in that way have an anti-inflammatory effect.

Each of these findings is a crucial puzzle piece that gives me a lot to think about. For example, the way people somehow look younger after a yoga lesson.

But I still haven't reached the end. I put up two new Post-it notes and write a word on each one.

The next morning, I get an email.

The message is as follows:

Hi Maria,

I realized that I forgot something very important.

We were talking about the role of bacteria in metabolism, and I might have downplayed it. That's mainly because I'm unsure about whether all the parts of the microbiota are involved in metabolism.

What I forgot is that among their most important roles is splicing complex sugars and dietary fibers into short fatty acids.

These are HDAC inhibitors . . . !

There's an old theory by Stuart Schreiber that the reason fiber protects against colon cancer is that HDAC inhibitors are formed from it. I believe this myself . . .

Have a nice weekend,
Tomas

This is interesting!

On one of my two notes on the bulletin board I've written *Good bacteria* and on the other, *Fibers*.

And the professor has just been thinking about both of them.

The H in HDAC, stands for histones, and HDAC is an enzyme that makes the DNA wind itself even more tightly around the little spool with its eight histones in the middle. An HDAC inhibitor does the opposite—it makes the DNA strand sit more loosely around the histones—which, as we saw earlier, produces strong health effects since the genes can "express themselves" more easily when they are looser.

Another image. If you squeeze someone extra hard, they have trouble breathing. If you let go, the person can breathe more easily. In the same way, an HDAC inhibitor can make the histones' arms relax around the DNA so that the genes can breathe and express themselves more easily.

These HDAC inhibitors have been known for a long time to stabilize the psyche.

Researchers, including Tomas Ekström himself, are now studying them in order to find new treatments for cancer and inflammatory diseases. In animal studies, people are looking at how they affect depression, and there are a number of ongoing clinical studies about HDAC inhibitors with the goal of finding new treatments for prostate, breast, and lymph tumors, for example.

We are forming a chain here: from bacteria and fibers to a demonstrable effect on inflammation, the psyche, and tumors.

The anti-inflammatory effect of probiotics might work through this kind of mechanism, where probiotics, to use plain language, chew their way through our food and leave their "poop" behind, which in turn affects our genes positively.

Now there are suddenly several possible partial explanations for the various anti-inflammatory principles. These explanations work through different pathways and fight inflammation by different means.

When the innovative researchers in Lund tried to change the health of their participants using an anti-inflammatory diet, might they have done so by activating several different mechanisms—at the same time?

In this detective work, I've gotten help from very knowledgeable scientists in digging and sorting through work at the absolute cutting edge. But no one has yet described the whole picture. No one can say yet that *this exactly* happened on the molecular level in all the participants in the Lund trial when all the different tricks were combined, or *exactly this* caused the epigenetic changes seen by Dr. Stellar's awe team in Toronto as the participants watched burning sunsets. Nor can anyone say that this is how it's all connected to the longevity of the Seventh-day Adventists in Loma Linda and the glowing skin of the women in Dr. Olsen's clinic in London. Or that *just like this* the histone arms in Professor Klarlund Pedersen are changed when she runs her laps around the bluish lakes in the center of Copenhagen.

No modern Western scientist has described this constellation as broadly as my Ayurveda doctor did a few years before, although then in different colors, with a different frame of reference, and with elegant, Indian English.

"Health, Madame, is about combining the right food, exercise, rest, massage, meditation, and yoga."

People already knew that in India three thousand years ago. So why has no one taken it further?

"You have to meet Jeya," says Tomas Ekström when I ask him this question.

"Who is that?"

But before I meet another person, other things get in the way. My sick friend is starting to lose ground and is much weaker. She lives in another city, but I want to see her.

What we talk about the last time we see each other will have to remain between us, but when she asks me to hold her, we meet in a long, heart-felt hug and tell each other "I love you." This farewell hug is as powerful as the welcoming hug you give your children when you see them for the very first time.

She looks at me with a gaze from eternity.

Now another vault is opening within you, and soon I won't be able to follow, I think with the utmost respect.

It's still too big and too hard to accept what this means and how sad I feel about all this happening. But I learn that bliss can be born out of circles that are closed. There's magic and awe in this completion, wholeness, sincerity, and richness, even if it's linked to great sorrow.

I keep talking via Skype to the coach who helped me in Los Angeles, and she gives me completely new perspectives when she blends the professional approach with spirituality. She supports me in accepting a number of difficult things for exactly what they are.

The coach says this:

"What if we aren't bodies that contain a little soul somewhere? What if instead we are souls who have a physical experience and are here to learn what we need to learn?"

Maybe it is so. All souls have their journey to make. I have mine. And what do I want that journey to be like?

I ask myself that question now, and the answer comes to me in a flash. As filled with light as possible—in thoughts, food, actions.

My lifestyle is part of this choice, and it has all become so self-evident to me over time that it's almost like an invisible, pinkish gas that whirls around and helps me quickly get back on my feet, even when I've lost myself totally. It doesn't protect me from the great waves of life, but it gives me more endurance and helps me attract others who can support me when I face challenges.

And my back, the problem that first made me start this journey?

Not a whisper of complaint. It feels strong and durable, as long as I follow my guidelines. If I get lost for a few days it starts to grumble.

My constant infections are also gone. When everyone else gets sick over the summer, I have one day when I feel off. I drink tea with turmeric, prepare my salmon and vegetables, take my probiotics and omega-3 tablets, drink smoothies made of spinach and protein powder, meditate and rest, and wake up the next day completely well.

My weight is stable, and I have more muscles than I did twenty years ago. I'm myself, but in a stronger form than when I began. And I'm beginning to get better at accepting things for what they are. Or as my friend told me:

"Sometimes you have to choose between being right and being happy."

Nobel Peace Prize for that insight.

Sometimes, but very rarely, it's absolutely imperative to be right, like when you're talking to dictators in Ethiopia or protesting the imprisonment of monks in Tibet.

But in human relationships, which are more complicated than the twisted DNA of millions of people put together, and in stressful work situations, you have to accept that we're all doing our best and that it might still be wrong. That in the big picture, it still doesn't mean as much as the love you can experience when circles are brought to a sacred completion.

I learn all this, and much more. It gives me the greatest feeling of awe, as does my meeting with Jeya Prakash.

So now I'm finally sitting with Jeya Prakash, a man whose appearance can best be described as a cross between Gandhi and ET. Jeya means "victory" and Prakash means "light"; I'm sitting with Victory Light at a breakfast restaurant in London.

He's the humblest doctor you could imagine.

"I hope I'm not delaying you from your very important airplane," he says apologetically in his Indian accent, since he knows I'm about to go on a trip.

"But I'm the one who wanted to meet you!"

"It's my pleasure to be able to meet you this delightful morning."

We exchange polite phrases, as you can only do with a charming Indian who shakes his head and smiles.

Jeya Prakash is a doctor on Harley Street in London. In the beginning, his focus was on cosmetic surgery and external beauty, but then he experienced a health crisis. By chance, Dr. Prakash walked into a body scanner at a medical conference and discovered that he had a bulging aorta, the artery that handles the vital blood supply from the heart. He had a so-called aortic aneurism, which is potentially fatal. After thus unexpectedly coming into contact with his own vulnerability and mortality, Dr. Prakash began to think about the beauty that comes from within—and aging.

But Jeya Prakash isn't only a conventional doctor of Western medicine. He grew up in Tamil Nadu in India, where he studied medicine in the capital, which he insists on calling by its old British colonial name, Madras.

"Madras—but I worked there for several years!" I say.

"You know where it is? Fantastic, then we have that in common too."

Yes, circles are closed. We talk about Tamil Nadu, about the beauty of the landscape, about the temples, and about the Tamils themselves, this ancient people in the southern part of India.

I thought that we were meeting to delve deeper into my questions about inflammation and aging. But life decides something else.

"I think Tomas thought we should meet because of the work I'm doing to try to bring Ayurveda into a Western framework and actually examine how it works."

I stare at him.

From Ayurveda to conventional Western medicine. Is this the man I've been looking for?

"You mean you're working on this?"

Like almost all Indians, Dr. Prakash was raised with Ayurveda, but then he studied conventional medicine and so came to treat aging in his profession, first cosmetically and then from the inside out. That was when he heard about the trials with starving rats who turned out to live longer than rats with access to food.

What lay behind all this, and how did the epigenetics work? he asked himself.

When he visited one of the medical temples in Tamil Nadu, a priest doctor sat with him, one of the priests who work at the temples and also have a medical role within Ayurveda.

"He asked me what my life task was," Dr. Prakash says.

Or his *dharma*, as the Indians call the life purpose we carry with us in our karma.

"I have to bring this knowledge to people; I felt it clearly inside myself. We are all born with our genes, our unique piano; but how we play that piano and what melodies we play, that's up to us. I needed to work with this. And I began to work differently."

Dr. Prakash believes that inflammation, along with a number of other mechanisms that also work together with inflammation (for example, methylation and oxidation), contributes to aging on all seven levels of the body: in the molecules, genes, cells, tissues, organs, systems, and the whole body. The doctors of his old homeland have been on the trail of this inflammation for a long time, just as I suspected, and tried to handle it through a holistic approach.

Dr. Prakash tells me about a wise old Ayurveda doctor he met.

"Your belly should be one-third full of food, one-third full of water, and one-third empty. Then you'll have a long life," he says.

That's precisely what the rat experiments showed, he realized—that food deprivation led to a longer life. So maybe there really was something to the ancient medical art of his homeland?

He decided to go through the healing principles and herbs of Ayurveda systematically to see if the effects could be quantified in the way that Western medicine requires in order to accept something as the truth. In this work, Dr. Prakash is collaborating with researchers from India and from Karolinska Institutet in Stockholm. He shows me long lists of substances that he wants to find out about, known substances from Ayurveda with names I recognize from my Indian spa, which he is now trying to link to certain physiological mechanisms.

Dr. Prakash and his colleagues are trying to close the circle between old Eastern empirical knowledge and newer Western conventional medicine, and one vault of human understanding after another opens up. But more research is needed.

In the course of my journey, there are several breakthroughs, like when scientists suddenly discover that chronic fatigue syndrome is caused by inflammation and that depression also can be linked to inflammation. Or when Professor Martin Schalling and other scientists from Sweden and Norway write in the journal *Läkartidningen* that antipsychotics of the future might be found among the new biological medications that have been developed to fight autoimmune diseases and that schizophrenia might be counteracted by new treatment strategies for inflammation. Or when researchers begin to look at whether ADHD might be ameliorated through probiotics, since they decrease inflammation, which has turned out to increase the risk of neuropsychiatric diseases.

Once again, one vault of new knowledge after another is opening up.

But we're still not quite at the end of this incredibly exciting journey. And when I say exciting, I'm thinking of the fascinating fact that we can be involved in shaping our own health, since we ourselves can control our diet, our exercise, our rest, and our awe, and make the most of the genes we have already. Or as Dr. Jeya Prakash would say, "Play the loveliest melodies on our piano."

That's why I'm now asking the collective scientific community, humbly but firmly, to take on this task on behalf of humanity, and in the course of doing so, to also include the study of ancient, empirical Eastern knowledge, approaching it with a sense of curiosity.

We want to know more about how all of this works, whether there are additional biochemical paths to explore, and which other factors affect inflammation—beyond diet, exercise, stillness, awe, yoga, sunsets, vegetable fibers, and God, and everything else that we've found.

We realize that here is an important key to dealing with and curing our society's ailments: cancer, cardiovascular disease, depression, and joint disease. We also realize that new techniques will arise in the borderland between gene technology and the amazing world of algorithms, and that these techniques will give us even more tools to work with.

So take the lead! We will listen.

But I am also saying, most humbly, that we can't wait until you're completely done. So while you're working, we'll take things into our own hands. We'll dig right here where we're standing, put together the knowledge that we do have, and put everything together to the best of our ability and creativity, to form a whole and a lifestyle.

Because you can live like this, and it gives you incredible benefits once you've made it all work.

The question is just exactly how to do it.

It's time to become very concrete.

"All souls have
their journey to make.
I have mine.
And what do I want that
journey to be like?"

"The anti-inflammatory journey is not a happy pill, but you do become happier and more mentally stable."

15. THE JOURNEY TO AN ANTI-INFLAMMATORY LIFESTYLE

Now I've done everything I said I would do in the beginning—I've undertaken a journey of knowledge through the anti-inflammatory landscape, where, as a science journalist, I've tried to go beyond the headlines and simplistic explanations. I've also tried to test everything myself and expose myself to success, failure, and learning.

Here is the conclusion that I've drawn:

We have scientific proof that low-degree systemic inflammation drives illness and that anti-inflammatory measures not only counteract disease but also strengthen everyday health in a powerful way. We see the evidence from the Blue Zones, from research in Lund, from the epigenetic front line, from awe research, and from a number of other things.

I can only summarize it like this: In my opinion, the anti-inflammatory journey is a journey toward your *best self*, a body and soul in balance, a feeling of harmony and alertness.

Let me also say this:

- The anti-inflammatory journey is not a happy pill, but you do become happier and more mentally stable.
- It isn't a doctor's visit, but central health factors are optimized.
- It isn't a diet, but body weight is stabilized and the body feels streamlined and strong.

- It isn't a beauty cure, but the skin gets new luster and strength and fewer wrinkles.
- It isn't an intelligence cure, but you sharpen your ability to think and your working memory.

We remain ourselves but in our very best version.

How can we achieve this? How can you put together an anti-inflammatory lifestyle?

Boost your system with anti-inflammatory foods. Every day, eat plenty of good, natural food, polyphenols, omega-3, and probiotics.

Lower sugar intake. Every day, spare the body from too much sugar and too many carbohydrates and reduce the glycemic response to the sugar that you do eat.

In motion. Give yourself the chance to exercise every day.

Stillness. Give yourself peace, calm, and conscious rest every day.

Seek out awe. Be curious about how to find your awe and allow yourself to experience great and divine moments.

Those are five points whose initials, through a small miracle, form the word BLISS.

The path to bliss is simple, but it requires a new consciousness. It involves a kind of updated version of pre-human Lucy's lifestyle, which deeply respects our human roots but also involves adaptation, since we no longer live on an African savanna but in a modern society with completely different frameworks. Through this lifestyle, we can find our way back to ourselves, although with new eyes.

Here I'll go over the principles that combine to form an anti-inflammatory lifestyle—the path to bliss.

THE BLISS PRINCIPLES

1. Boost with anti-inflammatory food
The value of real food. More polyphenols. More omega-3.
Increase probiotics.

Food is meant to provide health, joy, strength, and enjoyment. Just as Lucy did, choose food that nature has created, as close to its natural form as possible. In keeping with the slogan "made by nature, not by man," I'd rather eat a tomato than a prepared tomato sauce, rather an orange than juice, rather grilled meat and potatoes than prepared hash. Any pre-made, processed food with more than five ingredients should be regarded with suspicion.

Eat real, homemade food. Not weak salads that leave you vulnerable to a blood sugar crash in the afternoon, but "regular" food that's boosted with anti-inflammatory tricks.

Eat more vegetables of all kinds, preferably four different kinds with four different colors.

Eat a rainbow of vegetables and berries every day. Blueberries, purple eggplant, red onion, green spinach, yellow peppers, orange carrots, red tomatoes, and all other colors. Vegetables, with their various polyphenols, act directly or indirectly (researchers are investigating this) as protective mechanisms for the plants, and we humans can "borrow" their effects to protect ourselves.

Eat plenty of protein at every meal: poultry, eggs, lentils, meat, fish, or protein powder, which builds up cells, connective tissues, and muscles.

Eat plenty of fats, which give the body energy and enhance the taste of food. Oils like olive oil, rapeseed oil, coconut oil, avocado oil, and sometimes organic butter are good. But avoid margarine, sunflower seed oil, and hydrogenated vegetable fats in commercially prepared food.

Use spices, and use them with abandon; find new combinations. Mix and match; finding "spice families" that work together is fun and makes cooking easier. I like to combine thyme and garlic, turmeric and paprika, coriander and cumin, chili and mint, ginger and lemon. I am constantly looking for new ways to enhance food with spices that contain powerful polyphenols.

In general: add turmeric to everything! Our pots and pans at home are yellow from all the turmeric I use when I cook vegetables and when I sauté chicken, salmon, and steaks in coconut oil and turmeric to get a nice finish.

Love your tea. All kinds—black, green, red . . . Use a variety of different herbal teas every day.

Coffee contains polyphenols but also activates blood sugar. Compromise by having one cup a day.

Be careful with alcohol, but a glass of red wine can be taken since it (a) gives pleasure, and (b) contains the polyphenol resveratrol, which research has shown to be anti-inflammatory. Try to choose a red wine with a strong, slightly harsh taste, like pinot noir, which has the highest levels of resveratrol.

Eat omega-3 every day. Regardless of whether you eat fatty fish several times a week, take omega-3 in capsule form or get your daily dose from little chia seeds in a chia pudding that's been allowed to soak overnight in a glass of almond milk; you will soon notice how this fortifies everything from your mood to your skin.

If you're at a restaurant and don't know what to choose, go for fatty fish and vegetables. That's the new basic diet.

Grow your inner gut flora. (Hi there, bacteria!) Eat lots of greens, and also boost with a probiotic tablet every day, switching out the type of bacteria when you've used up the old jar. You want to expose yourself to many different kinds of good bacteria. Also eat extra yogurt or kefir every day. Experiment with kombucha, and choose some different flavors, like ginger, or turmeric.

I like to eat a mini bowl of fermented raw vegetables at dinner also, like

sauerkraut or kimchi, but realize that this might be too much of a gastro-
nomical boot camp for some people.

2. Lower sugar intake
**Keep sugar levels down. Eat better, complex carbohydrates.
Decrease the GI response.**

Carbohydrates are a complicated subject, I've learned. From gummy
raspberries to pasta carbonara—what's the best strategy?

There are two main goals: to decrease the amount of simple sugars
by eating better carbohydrates and in smaller amounts, and to moderate
how the body responds to sugar. This is to keep down the quick sugar
peaks that are the body's enemy since they directly drive inflammation.

Converted to everyday strategies, this means something like the
following:

Plan for the long term and even out satiety. And plan ahead so you
won't end up with panic and stress. The planning doesn't take more time,
but I've learned that it's another type of time, more "ahead time" than
"panic time." Which in turn gives better food quality.

A first step is to get rid of all the sugary junk in the refrigerator, freezer,
and pantry. Out with marmalade, ice cream, cookies, sodas, and such, so
that it will be harder to satisfy hunger with a sugar fix.

Another method is to figure out a new standard breakfast, which will
probably be different from the way you used to eat. Breakfast buffets with
bread, orange juice, sugary fruit yogurt, "regular" high-lactose milk, processed cereal, and marmalade—goodbye! Many of these products will
give you a blood sugar rush or create other types of inflammation. Bread,
even if it's whole grain, contains heavy gluten proteins that can give rise
to low-grade inflammation. (In my new existence, I have an occasional
slice of bread, maybe a Danish rye bread or sourdough bread, where bacteria have broken down some of the gluten proteins in advance.) Juice
contains as much as several teaspoons of sugar, without the fiber that

GOOD NUTS AND SEEDS

- Almond butter
- Almond milk
- Almonds
- Brazil nuts
- Chia seeds
- Hazelnut butter
- Hazelnuts
- Pecans
- Sesame sauce—tahini
- Sesame seeds
- Sunflower seeds
- Walnuts

naturally exists in fruit pulp and skins, lowering the sugar response when you eat whole fruits.

The new breakfast instead focuses on protein, fat, vegetables, and fruit. A smoothie with almond milk, fruit, nuts, and protein powder. A bowl of yogurt with nuts, seeds, and berries. Scrambled eggs, sliced tomatoes, cucumber, spinach, and rice cakes. Or a bowl of oatmeal with seeds and maybe an egg to keep protein and fat up.

The strategy is to choose fresh fruits and berries, and in the category of complex carbohydrates, unprocessed products like sweet potatoes, brown rice, quinoa, and oats are your friends. And eat the carbohydrates along with fat and protein!

I've decreased the amount of carbohydrates since I want to keep my insulin content low and even, but at the same time, body and brain need the energy that carbohydrates can give. Here you have to experiment to find the right level for you. One strategy might be to only eat complex carbohydrates in one bigger meal per day; for example, the one you eat right after exercising. In my case, that doesn't work if that meal is lunch. If I don't get complex carbohydrates at dinner, I can't sleep at night as the carbohydrates induce sleepiness.

I've also learned to eat the food on my plate a little differently. Now I always begin the meal with proteins, vegetables, and fats and eat the complex carbohydrates, like sweet potato and quinoa, in the later part of the meal. That makes the insulin level rise more gradual, and satiety is signaled via the foods that are least inflammation-driving. If a meal consists of chicken, grilled vegetables, salad, and brown rice, you should eat it in this order: first vegetables and chicken, then the brown rice at the end. I no longer eat a big plate of pasta Bolognese with a little sauce but have changed the proportions so that I have lots of vegetables, lots of meat sauce, and a smaller amount of pasta (preferably gluten free).

Read the list of contents and zoom in on the heading "carbohydrates," under which the sugar content is listed by itself as a subheading. Breakfast cereal with 25 percent sugar is candy; it is not food.

When it comes to lessening the glycemic response to a meal, there are two good, scientifically proven tricks: vinegar and soluble viscous fibers. You can take advantage of this by using vinegar in a salad before the meal,

YOUR BEST COMPLEX CARBOHYDRATES

Choose complex carbohydrates that have a low GI value and aren't inflammatory.

- Brown rice
- Buckwheat
- Gluten-free products—flour, cereal, bread
- Oats—gluten free
- Potatoes—once in a while, and with protein and fats
- Quinoa
- Sweet potatoes

If you have a craving for wheat-based bread, choose sourdough bread, which is less inflammatory since sourdough fermentation breaks down gluten. Or bake using gluten-free flours, like almond flour, buckwheat flour, or coconut flour.

NUTRITIOUS MILK PRODUCTS

From a purely evolutionary standpoint, it's odd that adult creatures ingest the milk of another species, like we human beings do. When we're little, we have the ability to process the lactose found in cow's milk and digest it, but most adults on earth can't drink cow's milk because the body (through epigenetic mechanisms) turns off the ability to digest lactose as we grow from children to adults.

We all have different levels of lactose tolerance, and you can be sensitive to it without being intolerant. At the same time, dairy products are tasty, protein-rich, and part of our culture. Again, feel your way to a level that's reasonable for you and your stomach.

In products like cheese or yogurt, the lactose is partly broken down and therefore easier to digest.

- Butter—organic
- Cheese—Parmesan, sheep's cheese like feta and pecorino, and goat cheese, preferably from animals that graze on some Mediterranean hillside where they eat anti-inflammatory herbs. (I eat good cheese a few times a week, and try to mix sheep, goat, and cow origin.)
- Cottage cheese—try it out, preferably lactose-free
- Crème fraîche—small amounts
- Milk—choose organic and lactose free if that makes you feel better
- Whole-milk yogurt—small amounts
- Yogurt—choose the full-fat Greek or Turkish variety

in the French style. Or take in more of the soluble fibers in vegetables, berries, beans, brown rice, figs, flaxseed, and sunflower seeds. These tricks will help you slow down your blood sugar increase after a meal, which is exactly what we want.

Snacks can be a challenge. Instead of cake and coffee, choose something more satisfying, like a bowl of Greek yogurt with chia seeds, two boiled eggs with a tomato, or a juicy red apple with nuts. Or try making exciting little energy balls at home in the mixer using coconut oil, dates, and berries, which you can take with you and pass out to fellow humans who may need it.

Personally, I'm skeptical about all the juicing. Pure fruit juice pushes insulin the way a race-car driver pushes down on the gas pedal, which ultimately drives inflammation. If you're going to juice, going green is best, preferably with some added nuts, whose fat gives a slower blood sugar rise. The very best smoothies are the ones that combine juice with protein and fat and, for example, contain nuts.

But if I get an irresistible sugar craving and eat half a box of chocolates a couple of times a year, I accept it with calmness. That's what happens if you put my inner Lucy next to a box of chocolates on a gray Thursday evening in November. Time to shrug your shoulders and continue the bliss cure.

3. In motion
Every day, get some kind of exercise. Every week, do some aerobic exercise, muscle-strengthening exercise, and stretching.

Every opportunity for exercise decreases inflammation in the body. You need to exercise every day, but you can vary the type of exercise, according to the rhythm of the day, your schedule, or the demands of your family, job, and different events in your life.

Part of the puzzle is getting in some kind of regular aerobic exercise every week, something that makes you sweat. Biking, power walking, running, skiing, swimming, tennis . . . You also need to do strength training, either in a class setting or by yourself at the gym, so that you can seriously challenge the muscles and build active muscle mass, which will help communicate with the immune system. Finally, add some calm, stretching, relaxing type of exercise. The body needs all three: aerobic exercises, muscle training, and stretching.

Many people talk about willpower, and I've come to see my workout sessions as my breathing, my freedom, something to enjoy. But you can't expect to always feel like exercising when you should. Exercise also needs to be planned, like everything else that's important. Exercise that only happens when there's spontaneous inspiration won't become a lifelong habit. Instead, aim to exercise a little every day, but vary the length and intensity. And even when you don't feel like it, you can at least try ten minutes. If it still feels hard, it may be time to take it easy. Usually, though, both motivation and enthusiasm will get you going, with their endorphins and dopamine.

You can also eat strategically for exercise. That means that you arrange carbohydrate intake around workout sessions, along with proteins, to make sure that the carbohydrates are used to build muscles. Preferably eat an hour after exercise—protein and carbohydrates—so that the hormone cortisol content will be decreased quickly. That will lessen the inflammatory effect of the cortisol, and your muscles will get nourishment more quickly.

4. Stillness
Actively seek out opportunities to de-stress:
meditation, yoga, mindfulness. Just be. Sleep.

The jigsaw puzzle of life is stressful for most of us, and that stress increases inflammation in the body through the activity and stress hormone cortisol. If you want to live in an anti-inflammatory way, you need to plan for rest just as much as for activity.

A moment of deep breathing and meditation decreases inflammation. So take the time to breathe deeply, meditate, practice yoga; to just be, to actively train mindfulness. Nothing really means as much as we think, except for the truly big values in life, like love.

It's also important to let the digestive system rest. Short periods of fasting decrease inflammation, and you might want to try some form of mini-fast every week. I don't like going all day without any food at all but have found that one or two days a week with 14:10 can feel very good. (That is, fourteen hours without food after the last meal of the evening. If I have dinner at seven I won't eat breakfast until nine the next day.)

This rest and stillness also includes healing sleep, which in itself contains important repair mechanisms. Maximize your anti-inflammatory sleep by focusing on keeping cortisol to a minimum, and try to fall asleep before eleven o'clock. That means giving yourself a little time to slow down before bedtime, turning off the computer and TV, and staying away from the internet and all social media. All these screens vie for our brains and keep us revved up when we should actually be slowing down.

5. Seek out Awe
Allow yourself to stop and enjoy the beauty of life.

Now we've come to the awe, based on the findings from the stellar research. Only you know what your awe looks like, but give yourself time to make a list of what feels big, holy, and beautiful to you. And give yourself permission to stop and take this in, to feel how your whole system slows down.

As I've worked on this book, people I've talked to have given me the following images that you might find inspiring:

- The most beautiful sunset over the ocean, where the sun becomes a glittering strip of light and you see sailboats out on the water.
- The untouched mountains with blue ski tracks, when the day comes to an end and you're alone on the tracks.

- The quiet church at Christmastime, with its lit candles and its calm devotion.
- Holding a newborn child.
- Being able to say thank you to life, even at times of large inner turmoil and sadness.
- The artists in the Bloomsbury group and the art they created.
- Being able to sit and look at decorating tips on Pinterest and enjoy different pictures of tiles, colors, and fabrics.
- The soft glowing light in a Rembrandt painting.
- The soundtrack to the *Lord of the Rings* films.
- Rachmaninoff's Second Piano Concerto.
- Listening to Queen sing "We Will Rock You."
- Helping a friend in need.
- A meeting at Amnesty where you save the world and fight for imprisoned dissenters along with like-minded people.

Find your own list, let it grow, use it, make a mood board with pictures, fill your Spotify list. Allow yourself to be passionate, look for your sacred spaces, and stop and take it all in.

Life is great. Thank you for letting me be here.

That's what a bliss day might look like.

But what if it feels like too much, to take in everything at once?

"Allow yourself
to be passionate,
look for your sacred
spaces, and stop
and take it all in."

Wisely and slow.

They stumble that run fast.

—William Shakespeare

16. SOFT START

For those of you who feel that you have neither the time nor the energy for big health revolutions right now but would like to test a somewhat softer variety, here's some inspiration.

The goal is to turn some small, strategic keys and bliss out just a little, like this:

- Eat a real anti-inflammatory breakfast and add more vegetables and protein-rich foods for lunch and dinner.
- Exercise three or four times a week.
- De-stress every day and let yourself experience more awe.

Breakfast is the big challenge, because our conventional breakfast is so heavily influenced by farming culture's basic products, which have entered the age of industrial food. Ready-made sandwiches, sweetened low-fat yogurt, sugary cereal, juice—that's just the kind of food you want to avoid with this lifestyle. Instead, you can use the first meal of the day to fill yourself with anti-inflammatory power that will work all day.

At the end of this chapter, you'll find some good breakfast recipes.

For lunch and dinner, you can eat the way you usually do and what you like, but change the proportions on your plate. Think about filling half of your plate with vegetables, preferably four different types. Think all the colors of the rainbow:

- Purple eggplant, red onions and radishes
- Red tomatoes and bell peppers
- Yellow bell peppers and turnips
- Orange carrots
- Green lettuce leaves, arugula, spinach, bell peppers, and broccoli

- White cauliflower and celery
- Blue blueberries and red onion

You can also increase the amount of protein, making sure that you're eating about a palm-sized amount of protein at every meal. With that simple strategy, you'll increase the amount of polyphenols in your diet and decrease your body's insulin response after every meal.

You can also let yourself be inspired by the long list of active and spicy inflammation-fighters and add as many as possible to each meal: capers, chili, basil, rosemary, oregano, turmeric, cinnamon, garlic, ginger.

Enjoy your food, add flavors, experiment, and chew slowly.

Also add regular exercise three or four times a week, for at least half an hour. You'll get the best results from a mixture of cardiovascular conditioning and strength training. A good combination might be jogging a couple of times a week, a good long walk, a gym session at lunch, and yoga at home. Or biking to work every day and lifting weights at home in the basement. Maybe you'll find exercise sessions to follow on YouTube or in an app. Pilates, swimming, jazz dance, or soccer, anyone?

In addition to this, you need daily de-stressing on a conscious level. Maybe look for a meditation app, practice mindfulness, go to yoga class, or spend time in nature. Let yourself breathe and find your peace.

Your daily awe is a matter of consciously seeking out the things that feel amazing and big to you. Wonderful music, art, spirituality or prayer, nature experiences, engaging in something bigger than yourself with other people—where will you find your bliss? Look for it actively and do the things that give you goose bumps. All of this decreases inflammation and is healing.

BREAKFAST RECIPES FOR AN ANTI-INFLAMMATORY SOFT START

Smart, Good-Looking Smoothies

A smoothie in the morning is a wonderful way to gradually wake up the entire body. But many recipes contain way too much fruit and too little protein. After a few hours, you're hungry again and your blood sugar goes up and down.

The idea is to combine good fat + protein + fruit.

The ultimate balance arises when you combine a small amount of polyphenol-rich berries, like blueberries, with good fats from almond milk, flaxseeds and nuts, and a large dose of protein. Blueberries kick-start the brain's chemistry and also have been shown to promote a feeling of joy and harmony. This smoothie has a mild and soft taste. You can also add a handful of spinach, which increases the anti-inflammatory effect.

¾ cup almond milk
1 handful blueberries
1 scoop protein powder
1 tablespoon flaxseeds

1 handful spinach
6 nuts (optional)
½ teaspoon cinnamon

1. Combine all the ingredients in a blender.
2. Blend.
3. Enjoy.

Classic with a Twist

Eggs warm you up and give you a feeling of satiety in the morning. Here's an enjoyable way to turbocharge your eggs. An important ingredient is the anti-inflammatory turmeric, heated in coconut oil. Pepper also increases the effect of the curcumin, the active substance in turmeric. You might also want to add some finely chopped chili peppers to your spinach.

coconut oil
a pinch turmeric (ground)
1 diced tomato
1 handful fresh spinach
2 eggs
salt and pepper

Serve with:

An apple or two rice cakes

1. Melt the coconut oil in a pan. Add turmeric until it hisses.
2. Place the diced tomato in the pan. After a while, you can also add spinach and a little pepper.
3. Place the warm vegetables on a plate.
4. Fry the two eggs in the same pan. Add salt and pepper.

Rita's Seed Bowl

For those of you who love cereal and muesli, the craving for your old breakfasts can be hard to overcome. Here's a mild, subtle breakfast bowl that Rita taught me, which is both satisfying and filling and reduces inflammation.

about ¾ cup almond milk or lactose-free plain yogurt
1 tablespoon chia seeds
1 tablespoon flaxseeds
1 tablespoon hemp seeds
1 tablespoon sunflower seeds
1 tablespoon almonds (chopped)

Serve with:

Fresh berries or 1–2 tablespoons of unsweetened lingonberry jam or applesauce

1. Pour the milk or yogurt into a bowl.
2. Stir in the chia seeds.
3. Let mixture stand in a cold place for 5 minutes.
4. Add the rest of the seeds and the almonds.

Sometimes when the November winds are roaring outside and I feel chilled to the bone, I make this seed bowl warm, like a kind of seed porridge, with 2–3 tablespoons of gluten-free oats, cinnamon, cardamom, and turmeric, which I warm up in water in a saucepan. Just when the porridge is ready, I stir in a raw egg to increase the protein content, and a little salt. Serve with almond milk or soymilk.

❦

You're much stronger than you think you are. Trust me.

—*Superman*

17. THE THREE-DAY CURE

For the impatient, the motivated, or the fast mover—here's a three-day bliss program. Feel your mood lifting, your belly flattening, and your skin becoming more lustrous.

In this program, you combine all the scientifically documented tricks for living in a maximum anti-inflammatory way, all at once. Because, as we discussed earlier, the effect will be the greatest when you combine all the different parts, or as the researchers would say: when the physiological strategies work in synergy.

You can also use this cure before an extra important day when you want to be your absolutely best self—in top shape for an important job situation, when you're meeting your future mother-in-law for the first time, or just before you're going to give a speech to two thousand people and inaugurate a Nobel party or two. Many Hollywood stars use similar cures before red-carpet events. You simply become your most *glowing* self.

These days will contain all the bliss steps in one boosted version. The result will be your own de-inflamed self.

In other words, you, only better.

- Boost with maximally nourishing food that reduces inflammation and activates anti-inflammation in a number of different ways.
- Lower sugar levels by avoiding refined sugar, eating fewer carbohydrates, and having a daily mini-fast.
- In motion—daily exercise and activity.
- Stillness—with daily breaks and active rest.
- Seek out awe—gratitude and that wow feeling.

BEFORE YOU BEGIN

If you live alone, it's probably easier to control your time. But if you have small children, it can be a challenge and might require some support from your partner.

One idea might be to do the program from Saturday to Monday, so you'll have time to get the hang of the rhythm during the weekend—sort of like a spa holiday for body and soul, although in a home environment.

To make it easier, do this before you get going:

- Food plan. How are you going to eat the five meals every day?
- 24-hour plan. When are you going to do the various activities?

THE PROGRAM
FROM MORNING TO NIGHT
When You Wake Up

Whether you wake up on your own or when the alarm rings, allow yourself to stay in bed and breathe for a moment.

- Breathe deeply, feel how your belly rises and falls, stretch your whole body like a lazy cat.
- Let go of your cell phone, don't check your email, and don't check Instagram, Facebook, Twitter, WhatsApp—they can wait.
- Think about what you're looking forward to today—where's your joy?
- Or meditate for ten to twenty minutes using any technique that works for you.
- Brush your teeth and go into the kitchen. Drink some warm lemon water with probiotics and take your omega-3 supplement. Take some time to write down your intentions for the day. How will you eat? When will you exercise? What are you grateful for today?

- Go for a short walk, fifteen to twenty minutes. Breathe, enjoy, listen to the most wonderful, awe-inspiring music you know.
- If you live near pretty nature, go there if possible. A fast morning walk is known to lower inflammation.

All of this works as a kind of mini-fast that decreases inflammation markers.

Breakfast

After you've slowly let your day begin, you can let the surrounding world come in via social media and email.

Breakfast on these days is a smoothie, full of anti-inflammatory power and with the right balance of carbohydrates, protein, and fat. It stabilizes the blood sugar, provides protein and good fat for long-term satiety, contains fibers that flush out the tubes, and includes powerful polyphenols in the form of blueberries and spinach that reduce inflammation.

Breakfast Smoothie

1 glass almond milk
1 scoop protein powder
1 scoop green powder
(spirulina or similar)
1 teaspoon cinnamon
(ground)
1 bunch fresh spinach

4 tablespoons berries
(raspberries, blueberries,
or blackberries)
3–6 nuts (Brazil nuts, walnuts,
almonds, or hazelnuts, etc.)
1 tablespoon flaxseeds

1. Blend all the ingredients together to make a smoothie. Enjoy with a cup of tea of your choice and a teaspoon of raw honey.

Late Morning and Afternoon

Keep your blood sugar stable during the day by eating small meals containing protein, fat, and vegetables or berries every time.

Eat one or two snacks depending on your level of hunger and activity. One of them should contain yogurt or kefir for the sake of the good bacteria.

Suggestions for snacks:

- 2 hardboiled eggs
- 1 whole tomato

Or

- 1 handful nuts
- 1 handful spinach leaves

Or

- ¾ cup whole-milk yogurt or kefir
- 2 tablespoons berries

Lunch and Dinner

Allow yourself to eat real, healing, and healthy food. Breathe properly as you eat; chew, and enjoy with all your senses.

Make at least one pretty meal per day. Light candles, play good music. The meals consist of:

- Fish, poultry, meat, an omelet, or vegetarian patties (these can be bought frozen at the grocery or health food store)
- Colorful vegetables—grilled or raw
- Dressing—made with good olive oil and vinegar
- Herbs—the more, the better
- A tablespoon of fermented vegetables
- At one of the meals, either lunch or dinner, you should also eat good-quality starchy food, like sweet potatoes, quinoa, brown rice, or beans of some kind

Begin by covering your plate with green leaves.

Build on it with fish, chicken, meat, omelet, or a veggie patty prepared by cooking, roasting, or frying in turmeric, coconut oil, and black pepper. One of the meals every day during this quick cure should contain a fatty fish, like salmon, sardines, mackerel, or herring.

Add more colorful vegetables and, once a day, a handful of sweet potatoes, brown rice, quinoa, or beans.

Add your favorite olive oil and balsamic vinegar and sprinkle herbs like basil, rosemary, and thyme on top; finish by adding chili and capers.

Add a tablespoon of lactic acid–fermented vegetables in one corner, which you eat first of all.

FISH AND SEAFOOD

If you don't know what to have for dinner, one sure anti-inflammatory tip is to choose a large, mixed-green salad including some of the following. Fish and seafood are among the most anti-inflammatory foods you can eat, and they give a light feeling of satiety that feels good.

- Anchovies
- Cod
- Crayfish
- Fish roe—from whitefish, carp, or cod
- Flounder
- Herring
- Lobster
- Mussels
- Octopus or squid
- Oysters
- Pickled herring—without too much sugar
- Salmon
- Sardines—fresh or canned
- Scallops
- Shrimp
- Tilapia
- Trout
- Tuna fish—canned or fresh

Drinks

- Water. Drink at least ten glasses per day and boost your water pitcher with anti-inflammatory herbs: basil, rosemary, mint, and lime or lemon.
- Tea. Drink as much as you want—black, green, red, and at least two cups of herbal tea per day.
- Coffee. A cup of good coffee per day is okay, and you should drink it within an hour of exercising so the sugar released by the caffeine can enhance muscle work.
- Kombucha. Fermented and yeasted green tea—drink a bottle a day if you feel like it. Check the sugar content.

Every Day

- Get moving. Choose the type of exercise based on how you feel. Maybe a group training session, lifting weights at home, yoga, or something else that you like. Exercise for thirty to sixty minutes, but not more. (If you have a cold you should take it easy, of course.)
- Experience awe. Look at something beautiful, listen to music that makes your heart sing, go to church, sing out loud, read a sensationally good poem, go to a museum, look at the starry sky or at your wonderful children, go to the theater, or watch a great movie. Let yourself feel "wow—this is big."
- Be close. If you have the opportunity, be close to someone you like very much in a way that fits into your life.
- Rest. Give yourself thirty to sixty minutes of de-stressing rest— perhaps an afternoon nap, meditation, or sitting by the fire.

All the Time

- Breathe calmly and deeply, using belly breathing. Every deep breath de-stresses.

Sleep

- Wind down and finish the day with a cup of calming herbal tea.
- Shut down all screens well before you go to bed. Take a nice bath instead, put on warm pajamas, read a good book.
- Try to get to bed by ten o'clock at the latest and turn out the light at eleven o'clock; that way you'll be most in sync with the body's natural desire to rest. If you stay up after eleven, the body will go into a different kind of breathing as cortisol levels begin to rise again. That makes it harder to fall asleep and to have a deep and calm night of sleep.
- Read something uplifting and inspiring.
- Say a prayer, or meditate.
- And gratitude again—think about five things that were good today. Feel how amazingly wonderful it is to be alive and to be a human being. Write down your thoughts if possible.

Good to Have at Home

- Here's what you need for a three-day program. You probably have much of it at home already, but don't worry if you don't have everything. Just switch it out with something similar—we aren't fanatics.

From the Bookstore

- An inspiring notebook. For gratitude, thoughts, and plans.

From the Grocery Store

Fruit and Vegetables

- Green leaves: spinach, arugula, and other types of salad
- Colorful vegetables: tomatoes, bell peppers, fresh chili peppers, broccoli, cauliflower, leeks, and so on. Think of the rainbow: something purple, red, orange, yellow, green.
- Berries

- Sweet potatoes
- Fresh herbs like basil, thyme, and mint
- Lemon or lime

Staples

- Brown rice or quinoa
- Turmeric (ground)
- Rosemary (dried)
- Black pepper
- Salt
- Olive oil—choose extra virgin
- Vinegar
- Nuts (hazelnuts, almonds, Brazil nuts, or walnuts)
- Seeds (sunflower seeds, chia seeds, hemp seeds, or flaxseeds)
- Tea—black, green, or red
- Coffee

Protein

- Salmon fillets
- Chicken fillets or pieces of meat
- Eggs

In the Dairy Section

- Almond milk
- Greek yogurt (whole milk)
- Milk for coffee and tea, lactose-free or soymilk

From the Health Food Store or Pharmacy

- Protein powder (If you can tolerate milk, you can buy a powder based on whey; otherwise choose a vegan alternative. Just make sure the sugar content isn't too high.)
- Green powder, like spirulina
- Coconut oil, raw or virgin
- Honey, as fresh as possible

- Herbal tea, preferably several kinds (for example, peppermint or chamomile)
- Omega-3 supplements
- Probiotic (make sure it includes some form of lactobacillus.)

WHY IT WORKS

This lifestyle combines the best insights from research about inflammation and bombards the body and mind from a number of different angles, all in order to optimize physical and mental health.

- The anti-inflammatory food keeps sugar levels down, as well as the insulin response that drives low-degree inflammation.
- The big intake of protein gives a feeling of satiety, decreases the glycemic index of a given meal and hence the insulin response, and repairs the body.
- The colorful vegetables, berries, spices, and beans contain many anti-inflammatory polyphenols and soluble viscous fibers that are needed for the good bacteria in the gut.
- The high intake of omega-3 fatty acids from fatty fish, nuts, and seeds boosts the genes in an optimal way.
- Probiotics from supplements and fermented vegetables increase the gut flora and so contribute to decreasing cytokines, the inflammation markers that depress your mood and cause anxiety and a general feeling of discomfort.
- Awe (gratitude, wow feeling, beauty, nature, spirituality) decreases the level of inflammation markers.
- Exercise lowers blood sugar and reduces inflammation.
- Rest decreases stress, which reduces the risk of low-degree inflammation.
- Mini-fasts reduce inflammation.
- Closeness increases the oxytocin content, which in turn reduces inflammation.

✤

"Sometimes I want food that's
comforting and simple,
sometimes glamorous
and festive."

18. MY NEW LIFE IN THE KITCHEN

For me, the journey toward an anti-inflammatory lifestyle has brought new enjoyment in the kitchen, where new flavors and new life forces come together.

I like food that's nice looking, smart, and flavorful. Sometimes I want food that's comforting and simple, sometimes glamorous and festive. Since I began this transformation, I've become much more experimental in the kitchen. I've moved away from the bland consistency and taste of "white" food toward more adventurous, richer tastes, more varied and stronger colors. I can't get enough of testing new combinations of the different anti-inflammatory ingredients.

I'm in no way a trained chef but rather a home cook who likes to make food for all the nuances of real life, from gray foggy breakfasts to fancy Sunday dinners with the family, Fridays with close friends, and relaxing by the fire. My husband is an equally engaged cook, and we like to experiment with new and old flavors.

Here are some of the recipes that give me life force, taste, and joy. I've freely borrowed from the different food traditions of the world and exchanged recipes with friends, which I've adapted and modified over time. All the recipes have one thing in common: they boost anti-inflammation and, at the same time, provide wonderful flavors that you can enjoy in daily life.

BREAKFAST, LUNCH, DINNER, AND SNACKS

Jessica's Pancakes

MAKES 4–5

My friend Jessica lives in Oslo, and a few years ago she developed problems with her joints. She began to change her diet and sought out healing recipes. Here are her inviting anti-inflammatory pancakes, mild and comforting for breakfast or after a hard workout.

1 half apple
2–3 eggs
½ cup gluten-free oats
1 tablespoon flaxseeds
a pinch or two of salt
coconut oil, for frying

1 teaspoon turmeric (ground)
2 teaspoons cinnamon (ground)
1 teaspoon cardamom (ground)

1. Grate the apple.
2. Mix it with eggs, oats, and flaxseeds. Add salt.
3. Heat coconut oil in a frying pan.
4. Sprinkle turmeric, cinnamon, and cardamom on top of the oil and let it sizzle for a minute or two.
5. Pour the batter into the pan, a spoonful at a time, and build up the blobs to make bigger pancakes.

The apple can also be replaced by berries, for example, blueberries or raspberries. I recently tried pear chunks instead of grated apple and fried the pancakes in oregano, turmeric, and cinnamon, which provided a rich and sweet background to the elegant pears.

Serve with some spinach leaves and unsweetened lingonberry jam. This provides a long-lasting satiety, which feels good.

"Bread"

I was inspired by a bread idea from master chef Jonas Lundgren, author of the Paleo cookbook *Jonas paleokök*. This "bread" will fool even the most adamant gluten fans.

10 tablespoons coconut oil (plus a little extra for greasing the pan)
Sunflower seeds for lining the pan
2 cups nuts (hazelnuts, almonds, Brazil nuts, etc.)
2½ cups seeds (flaxseeds, chia, sunflower, etc.)
about 2 heaping teaspoons bread spices (ground)
⅓ cup rose hip flour
¾ cup dried fruit and berries (prunes, lingonberries, cranberries, apricots, etc.)
8 eggs

1. Melt the coconut oil in a saucepan.
2. Grease a loaf pan with coconut oil and cover the bottom with sunflower seeds.
3. Stir nuts, seeds, spices, and rose hip flour in a bowl.
4. Chop the dried fruit and blend it into the mixture.
5. Add cooled coconut oil to the dry mixture and stir to combine.
6. Crack the eggs into the bowl and continue stirring the batter, which will now become clumpy and gooey.
7. Press the batter into the loaf pan with your fingers until it feels even and then bake in oven. It will take about 50 minutes at 300 degrees F.

Cut the bread into thin slices and serve with smoked salmon and an omelet, or as a sandwich with almond butter, sliced tomato, and salt. Yum. This combo, rich in protein, omega-3, and fruit, is perfect before exercising. Two hearty slices of this bread along with a cup of Lisa's Bulletproof Coffee (see page 322), and you'll have enough energy to move to perform miracles in the gym or elsewhere.

African Curry

SERVES 6–10

This is a cool curry with roots in African history. In the late nineteenth century, Indian guest workers traveled across the ocean, from Gujarat in India to Kenya in East Africa, to help build the railroad that would go from the old trading town Mombasa, by the sea, via Nairobi in the highlands, and then wend its way toward Uganda and Lake Victoria.

There are stories about lions who would slink along the railway and eat the guest workers who didn't know how to protect themselves from these wild animals. But the railway stands today as a monument to their work. So does African curry, originally an Indian curry with African touches that has also borrowed flavors from Arabian cuisine. It's rounder, softer, more complex in taste, color, and form.

This recipe is from southern Kenya, near the border of Tanzania. A Swedish family runs a guesthouse and hotel there, near the blue ocean. Every Sunday, the chef serves a curry on the lawn and brings it to the windswept wooden tables. The chef has generously shared his recipe, which I've taken the liberty of changing to include a number of other active ingredients.

Hot, mild, sweet, strong, sour, bitter, and peppery. Everyone can create their own curry with a personal mix of ingredients. I like to prepare this one for my friends on my birthday. It's a guaranteed social taste experience when everyone compares their different plates and samples their way through the flavors.

Condiments to serve with curry.

3 lbs chicken thighs

4½ lbs tomatoes

1 lb onions

6 cloves garlic

coconut oil for cooking

1 tablespoon turmeric (ground)

1 tablespoon curry powder

2 tablespoons spicy cumin (ground)

about 2 inches ginger (fresh)

2 cinnamon sticks

1 tablespoon cardamom (ground)

3 bay leaves

2 tablespoons tomato puree

salt and pepper

1. Divide the chicken thighs into smaller pieces.
2. Place tomatoes in a saucepan, cover with water, and bring to a boil. The tomatoes are ready when the skin comes off. Peel off the skin and coarsely chop the tomatoes into large pieces.
3. Slice the onions in thin slices.
4. Finely chop garlic.
5. Heat the coconut oil in a large saucepan. Add turmeric, curry powder, and onion and sauté until onion is golden brown.
6. Add all other spices and tomato puree to the onion mixture and mix thoroughly.
7. Add the chopped tomatoes.
8. Add salt and pepper to taste.
9. Let this tomato and onion sauce simmer over low heat while you finish cooking the chicken thighs.
10. Heat coconut oil in a frying pan and fry the chicken pieces until they are just cooked through.
11. Place the chicken on top of the tomato and curry mixture, and pour the frying fat from the chicken over it. Let everything cook together for 2–3 minutes.
12. Remove ginger pieces, cinnamon sticks, and bay leaves before serving.

Time for your inner artist to make an entrance! Serve this curry along with brown rice (preferably cooked together with cinnamon sticks and star anise) and four to eight different condiments, elegantly served in little bowls. Choose additions that vary in color and intensity and add shape and contrast. Something yellow, something green, something cool, something hot, something crispy, something oily.

Here are some suggestions:
- Chopped chili
- Chopped egg white
- Greek yogurt
- Cubed mango
- Chopped peppers
- Chopped cucumber
- Cut coriander
- Snipped parsley
- Thinly sliced banana
- Thinly sliced pineapple
- Mango chutney
- Coconut flakes
- Chopped almonds
- Fresh spinach
- Lime wedges
- Chili oil (take regular olive oil, add some strips of chili, and let it stand for a while)

Serve the curry on the stove, and place the rice and the bowls of little additions next to it in a long row. Inspire your guests to combine their very own curry.

If I have guests who are vegetarians, I set aside some of the tomato and curry mixture and add cooked green beans, broccoli florets, and cooked carrot slices to make a vegetarian version.

Veggie Box

I have a dear old friend who's a chef and a philosopher of life. He taught me that "if you mix everything together, everything will have the same taste." That made me think about my plates of roasted vegetables, where I would mix vegetables, oils, and spices over and over again. It was good, but everything tasted the same.

That's how my "Veggie Box" was born. It's a beautiful artwork of tastes, where a lot of different vegetables are roasted at the same time in the same dish, but not necessarily at the same pace or with the same spices. This might sound like a time-demanding project for a crafter, but in fact it's a quick and relaxing thing to make.

I line an ovenproof dish with parchment paper so that it sticks up over the edge, nice and even. Inside, I make neat rows of vegetables and herbs, full of anti-inflammatory substances. Each row is made up of one vegetable. It becomes more interesting to eat since each row has its own color, shape, and taste, but the different tastes still melt together in some magical way. I'm not fussy about making the rows too precise but paint with broad strokes when I do this. It tastes heavenly anyway.

And the dish goes with everything—vegetarian patties, chicken, game, lamb, omelets.

Suggestions for "rows" in a Veggie Box:
- Thinly sliced carrot sticks with minced ginger and parsley
- Thinly sliced red onion with halved cherry tomatoes, oregano, and black pepper
- Yellow peppers in strips with turmeric, cumin, and coriander
- Zucchini in half-moon slices with chili and mint
- Brussels sprouts sliced in quarters with rosemary, dried or fresh

A Veggie Box using ingredients that I had on hand. A row of tomatoes and thyme, a row of padrone peppers and cumin, zucchini with my favorite mix of chili, mint, and sliced bell peppers with dill and turmeric. What do you have at home?

- Parsnips sliced in thin sticks with grated lemon peel
- Tomato wedges with thyme and garlic or dill
- Chopped white cabbage with a little cumin and rose pepper

1. Plan your rows beforehand so that you'll have a good mix of colors.
2. Arrange the rows—try to do it with some feeling and creativity.
3. Dribble olive oil on top and sprinkle a little salt on the finished masterpiece.
4. Into the oven, at 350–375 degrees F, for 20–30 minutes, depending on your oven and whether you want a little crispness in the vegetables or you like them well roasted with a lightly browned surface, like I do.

ANTI-INFLAMMATORY SAUCES

Since I'm (a) a little lazy and (b) a sauce lover, I discovered early on that you can actually "pour" anti-inflammation onto your food, in the form of sauces with a high level of anti-inflammatory substances.

I almost always have two good sauces in the refrigerator that I can use to transform everyday food by adding varying degrees of spiciness and acidity and that work with quinoa, chicken, fish, sweet potatoes, and fresh vegetables.

Creative Pesto (or Gustaf's Sauce)

First, a pesto sauce. My brother-in-law likes to say that every family should always have a good pesto in the fridge. Worth considering!

Once, my husband happened to throw away my newly prepared pesto in the course of an impromptu cleanup around the sink. It was sitting in a bowl, and he thought it was some kind of leftover. I didn't have any more basil at home, nor did I have Parmesan or pine nuts, and I got a little upset. At that point my son Gustaf calmly suggested that instead of getting all worked up, we should use the parsley and thyme that we had on the kitchen windowsill. I found some walnuts and a piece of hard, tangy Swedish Västerbotten cheese in the fridge. It was a success.

The takeaway is that we need our children more than we think—and that even a pesto can be created more freely.

As it turns out, basil can be replaced with other green herbs. Parsley and thyme are highly anti-inflammatory, as are rosemary, oregano, sage, and lemon verbena.

Pine nuts, a traditional ingredient, are expensive and can be replaced by other nuts, like walnuts (deeply anti-inflammatory) or hazelnuts. Or almonds (which, strictly speaking, are stone fruits and not nuts in biological terms but work anyway). Seeds, like sunflower seeds, also work well.

Parmesan cheese can be replaced with other hard cheese like Västerbotten cheese, cheddar, or Gruyère. I like my pesto with a coarse texture, and I both grind and serve it in the mortar we inherited from my mother-in-law. It adds some feeling.

Play around with different ingredients, starting from the following basic recipe:

1 pot fresh herb of your choice
3–6 cloves garlic, peeled
1 handful nuts of your choice
¾- to 1-inch piece hard cheese of your choice
½ to ¾ cup oil of your choice
½ lemon (can be replaced with white wine vinegar or balsamic vinegar)
salt, optional

1. Crush the herbs in a mortar.
2. Add the peeled garlic and crush it as well.
3. Crush the nuts in the same mortar.
4. Shred the cheese into small flakes and mix.
5. Mix everything together and drizzle with oil.
6. Add lemon juice or other acid to taste—and maybe a little salt?

Use the pesto:
- On a mussel omelet
- With grilled chicken thighs
- With a piece of grilled salmon or tuna
- With cooked fish

- With grilled pork chops or lamb
- With vegetarian patties
- With grilled sweet potato wedges

Harissa

Harissa is originally a hot, smoky paste from North Africa in which the sweet taste of roasted red pepper contrasts with fresh lemon and hot spices. My harissa is a little softer and milder than the original Tunisian version, more of a warm vegetable sauce than a hot paste. My family often asks for harissa, and that's the best report card. This recipe contains the sensational and sensual mix of cumin, coriander, turmeric, paprika, and chili, which is so characteristic of North African food. The combination of turmeric, oil, and pepper is an explosive anti-inflammatory mixture.

I make it according to the mantra 6–3–3–1, to keep the proportions balanced between sun-dried tomato, red pepper, garlic, and chili.

3 red peppers, in chunks

3 cloves garlic, peeled

1 whole fresh chili pepper (Choose the type of pepper according to how much heat you want.)

about ½ cup olive oil

½ tablespoon turmeric

1 tablespoon cumin (ground or whole seeds)

½ tablespoon coriander (dried, not fresh)

½ tablespoon paprika powder

a few cloves (whole, dried)

6 sun-dried tomatoes

juice of one half to a whole lemon, depending on how much acid you want

black pepper

1. Place the pepper pieces, peeled garlic cloves, and whole chili pepper on a baking sheet. Roast in the oven at 400 degrees F

for 35–40 minutes. The peppers should be brown at the edges and the peels should begin to peel off the pulp.

2. While peppers are roasting, prepare the flavorful spiced oil. Pour olive oil into a frying pan and add turmeric, cumin, coriander, paprika powder, and cloves. Let the spices cook gently in the oil until the flavors are released, about 5 minutes.

3. Let the oil cool.

4. Place the oven-roasted pepper pieces, chili pepper, and garlic in a mixer, along with the sun-dried tomatoes. Pour the spiced oil on top and blend everything until it has a slightly coarse consistency.

5. Add lemon and black pepper to taste.

Serve on the side with:

- White fish, like cod or sole
- Fried eggs, for breakfast or lunch
- Roasted chicken
- Various kinds of meat: lamb, game, beef
- Salads, like a salsa

Sauce Xipister

This recipe is inspired by the cookbook *My Little French Kitchen* by Rachel Khoo, who taught the British how to make traditional French home-cooked dishes from the smallest kitchen in Paris.

Xipister is pronounced "chippister." It's a lovely basic sauce from Basque, the borderland between Spain and France, which mixes the fragrant herbs of the region and lifts anything grilled to new heights. It works just as well for roasting vegetables in the oven.

I like to have a bottle of this sauce on hand to pour over grilled fish or chicken or to use as a grilling marinade or dressing for vegetables.

about ½ cup olive oil

about 1¼ cups cider vinegar

1 whole fresh chili

1 tablespoon capers

2 cloves garlic

1 dried bay leaf

1 sprig rosemary (fresh)

1 tablespoon thyme (dried)

zest of one lemon

1. Place all the ingredients in a clean bottle or jar. Close the lid.
2. Shake well and let rest in a cool place for about a week.
3. Done!

Caper and Dill Sauce

Dill is my favorite. It tastes like home, like the dill and meat of my childhood, and midsummers with my grandfather in the Dalarna region of central Sweden. When I'm in London, I always long for dill. Besides, dill is anti-inflammatory, as are capers. Here is a sauce that's inviting and also has a bit of a bite.

1 part white wine vinegar

3 parts olive oil

2 tablespoons minced capers

2 tablespoons chopped fresh dill

zest of one lemon

salt

black pepper

1. Mix vinegar and oil.
2. Mix in the capers, dill, and lemon zest.
3. Add salt and pepper to taste.

Good with:

- Grilled fish
- Grilled chicken
- Grilled vegetables
- Salad with eggs and shrimp—as a flavorful dressing

Emily's Tahini

My digital training buddy Emily Johns is a geologist and a vegan, and she makes amazing food. She named her blog and Instagram account after her favorite stone and her favorite vegan food: quartzandquinoa. I recommend it to anyone who wants some vegan inspiration with a high degree of innovation. When we met in Canada, she offered me a salad with "double dressing," as she called it, which was a new taste sensation for me. It's simply tahini mixed with vinaigrette.

Tahini has been a staple in my kitchen ever since. I whip it up when I need some protection from the noise of the big world outside. Its creamy, sesame-textured, dark undertone contrasts delightfully with the garlic and lemon.

Tahini is the base of many other sauces, including baba ganoush with roasted eggplant, lemon, garlic, cumin, and olive oil. I've even had a dream about this sauce—that's how good it is!

about 1 cup tahini (sesame paste)
½–¾ cup lukewarm water
4 cloves garlic
2 teaspoons spicy cumin
½ teaspoon chili pepper (dried)
1 teaspoon minced parsley (fresh or dried)
juice of one lemon
salt

1. Mix all the ingredients in a blender. Add water until the sauce is just the right consistency; it will thicken fast.
2. Serve with chicken, lamb, omelet, or cooked or grilled vegetables. Or use it in a salad, along with a regular vinaigrette, like Emily Johns does.

DRINKS

Lisa's Bulletproof Coffee

The basic recipe for this drink of freshly brewed coffee, coconut oil, and organic butter was created in California and contains medium-chain triglyceride fats that are easily digested, as well as caffeine, which gives an extra kick. Cold-pressed coconut oil has been shown in animal studies to be anti-inflammatory, and it's thought that the active component, so-called lauric acid, is equally effective in people.

I was inspired by my hairstylist, Lisa Daly, who not only has been taking care of my hair for the last eight years but also supplies me with life wisdom, laughter, and good vitamins; she has become my friend on the anti-inflammatory journey. She's also created a sensationally good version of bulletproof coffee. The result is a creamy taste sensation with dark and spicy undertones.

coffee (either a mug of brewed coffee or 1–2 espresso shots mixed with a cup of warm water)
½ teaspoon cinnamon
½ teaspoon turmeric (dried and ground)
1 pinch chili powder

about ½-inch piece ginger (fresh)
1 teaspoon coconut oil
1 teaspoon butter (organic and unsalted)
1 square dark chocolate (85%)

1. Brew the coffee.
2. Add the spices.
3. Add the coconut oil and butter, then add chocolate.
4. Zoom—mix in the blender.
5. Done!

Healing Water

How can you have a recipe for water? The original idea comes from regular lemon water, but you can mix up tasty new varieties that give you an anti-inflammatory boost and add a little beauty and glamor to your more sober life.

So here goes: Fill a glass carafe with tap or mineral water and add, for example:

- Blueberries and some thyme sprigs
- Raspberries and a couple of rosemary sprigs
- Lemon, lime, and tangerine
- Basil and lemon
- Lemon and parsley sprigs
- Sliced strawberries and some mint sprigs

Can also be mixed with kombucha for a bubbly and refreshing drink.

Greger's Morning Tea

The day begins early and wordlessly at our place, with candlelight and tea. My husband makes a smoky tea mix that's addictive.

1 package Lady Grey (loose-leaf tea)

1 package Lapsang souchong (loose-leaf tea)

1. Mix the two types of tea leaves thoroughly and store them in a jar with a tight-fitting lid.

Blue Zone Tea

The Blue Zones inspire us to enjoy nature's leaves and herbs brewed in warm water. That's what the people who live the longest do all over the world—drink green tea, tea made of herbs and leaves. Imitate the Blue Zones by making tea elegantly and simply using:

- Some sprigs of rosemary and hot water
- Some basil leaves and hot water
- Minced ginger, raw honey, some lemon slices, and hot water
- Green matcha tea, a powder that you dissolve in hot water
- Dried thyme, marjoram, oregano, and/or sage in a tea ball strainer, plus hot water.
- Green tea brewed with cold water

"Cold-brewed green tea with rosemary tastes incredibly fresh and takes only 10 seconds to make."

Emily's Great-Skin Tea

My friend Emily Johns makes a wild and fantastic rooibos tea. Three days of drinking this brew and your complexion will get a new vitality. It's highly anti-inflammatory, and I often use it when someone in the family is getting sick, since it also seems to put the brakes on a cold.

½ teaspoon cinnamon
a few cloves
½ teaspoon turmeric (ground)
1 pinch chili powder
about a 1-inch piece ginger (fresh)
1 tea bag rooibos tea
1 teaspoon raw honey

1. Boil 6 to 8 ounces of water in a saucepan and add the dried herbs.
2. Mince the ginger and place in a cup. Add the spiced water.
3. Place the tea bag in the cup.
4. Let steep for 5–10 minutes.
5. Remove the tea bag and strain out the spices.
6. Sweeten with honey and add some almond milk if you like.

MORE TEA IDEAS

My beloved grandmother drank eight to ten cups of tea per day—elegant and fresh in the morning and increasingly smoky as the day went on. She died at the age of ninety-six, vital to the end in every way. Tea is not only a wonderful drink but is also rich in polyphenols.

Power Matcha

When I'm writing long reports and lose my drive and inspiration, I usually feel like eating a sweet roll with coffee, something to pick me up. That's when this drink is perfect.

1 mug almond milk

1 teaspoon cinnamon

1 pinch chili powder

1 teaspoon cardamom

1 teaspoon turmeric

1 heaping teaspoon matcha (green tea in powdered form)

1. Boil the almond milk with the spices. Let it simmer for a minute or so.
2. Add the matcha powder with a special matcha whisk or a regular fork.

THIS IS WHAT I AVOID

What keeps me going is not denial, but desire and possibility. That's why, for the longest time, I avoided making a list of things to avoid. But people often ask what not to do, so here are the things that I've cut back on during my anti-inflammatory journey, with explanations and suggestions for things to do instead. Read the table that follows and see how it relates to you. What makes your belly swell up? This is often a sign of low-grade inflammation. What makes you feel light and energetic? That's where you'll find your best food.

Note: If you have any kind of pronounced sensitivity to specific food substances, for example, celiac disease, a milk protein allergy, or a nut allergy, you should of course follow your doctor's recommendations.

I'm cutting back on	because	and instead you can
Bread baked with wheat, rye, and barley	Gluten rich, see under Pasta. When the gluten effect of the bread is also combined with refined flour, which has a high glycemic index, the combined effect of gluten and high GI drives inflammation even more.	Switch to rice cakes! Nowadays I eat bread only a few times a week. Then it's either sourdough, where the gluten molecules are partially broken down, or a good rye bread with lots of whole grains, since these contain a lot of nutrients and lower GI value. In that way, you avoid the cumulative GI and gluten effect. (Note that barley bread baked with whole barley seeds that have been allowed to swell can have an anti-inflammatory effect.)
Cereal, muesli, and granola	High sugar and gluten content, which when combined have been shown to be extra inflammation causing.	Make your own "seed bowl" with nuts, seeds, fresh fruit, some berries, and almond milk. A little cinnamon on top lowers the GI levels, as does the oat cereal called Betavivo, which you can buy in health food stores.
Pasta	White wheat flour contains gluten, a hard-to-digest protein that can be linked to inflammation—not only in people who are gluten intolerant but in others as well. Early humans did not naturally ingest large amounts of gluten.	Use sweet potato, quinoa, and brown rice as complex carbohydrates along with your meat sauce. If I'm invited out and pasta is served, I'll take a small portion of pasta and more sauce and side dishes, like vegetables, Parmesan, parsley, and so on.
Pizza	Gluten rich.	Make an omelet with pizza toppings like tomatoes, Parmesan, onion, tuna, olives, etc.
Hamburgers	The buns are gluten rich.	Eat half of the bun like an open-faced sandwich with meat, lettuce, and tomato on top. But this probably shouldn't be an everyday lunch.

I'm cutting back on	because	and instead you can
Prepared soups	Read the ingredient list and watch out for high sugar content and too many ingredients. These often contain wheat flour as a thickening agent.	Choose carefully. A prepared tomato and lentil soup can be fine. Otherwise you can easily put together a wonderful soup using cooked vegetables, vegetable broth, and almond milk, finished with a handful of almonds and a little salt. Good to make from root vegetables like celery, carrots, and onion.
Commercially produced cookies	They contain trans fats, gluten, and sugar. Trans fats are hydrogenated vegetable oils that have been shown to be extremely inflammatory; the same is true for sugar. Gluten also drives inflammation. Together—sorry, I say this with sorrow in my heart—it's a triple inflammation bomb.	Bake your own cookies using coconut oil or good organic butter, honey, eggs, gluten-free oats, almond milk, nuts, and small amounts of dried fruit. Or make tasty "raw energy balls" out of coconut oil, dates, and seeds! You can also use sweet potatoes, carrots, etc. in cookies and sweet breads. I sometimes add a little protein powder, but that may be too hard-core for some people. There are a lot of recipes on the web if you're looking for gluten-free and lactose-free recipes.
Candy	Candy contains sugar, which drives inflammation—period. Sometimes it also contains trans fats, stabilized vegetable fats that also drive inflammation. These are in the process of being banned but are still found in grocery stores.	Eat good-quality dark chocolate, preferably with more than 70% cacao content. Cacao beans are rich in anti-inflammatory polyphenols. It's often enough to have two or three squares, since the taste is so rich and complex.
Chips	Chips contain trans fats that drive inflammation.	Replace with olives, smoked almonds, some slices of cucumber, some slices of good ham, carrot sticks with hummus. With a glass of wine, it gives the same festive feeling but with a completely different nutrient profile.

I'm cutting back on	because	and instead you can
Dairy products	Dairy products contain lactose (milk sugar), which is broken down by the lactase enzyme. People who have low levels of lactase can become gassy and bloated from drinking milk. The link between inflammation and lactose is still unclear. It's clear that people with a manifested lactose intolerance develop a measurable low-grade inflammation from dairy products, but for people without lactose intolerance, research doesn't give any clear answer about the inflammation link. As for fermented milk products, like yogurt and kefir, studies have shown that they have a definite anti-inflammatory effect.	Experiment to see how much milk you feel good ingesting and vary between lactose-free and other products. Dairy products are good sources of calcium, and some of them also contain high levels of good bacteria. Organic milk from grass-fed cows also has a higher omega-3 content. My compromise: • A little regular milk, preferably organic, in coffee and tea. • Lactose-free yogurt or a whole-milk Greek variety, five times a week. The rest of the time I use almond and soy milk.
Ice cream	It contains sugar, lactose, and sometimes the wrong kind of fats. Sorry, but this is not part of my diet.	Make your own ice cream. If I crave the sweetness of ice cream, a frozen banana mixed with some fresh strawberries can be a good dessert sorbet. If I long for the creaminess, nut butter spread on apple slices can be good. If you want real ice cream, choose a kind that's made of good cream.
Coffee and energy drinks	High levels of caffeine drive up the stress hormone cortisol, which raises the blood sugar level and triggers inflammation processes in the cells. At the same time, coffee contains good polyphenols, which have qualities that protect the brain against Parkinson's and Alzheimer's, among other things.	Draw the line after one or two cups of coffee per day. Choose a good-quality coffee, and preferably drink it before exercising so that the sugar that's released in the blood when the caffeine sets cortisol in motion is used to give energy and build muscles. I also drink a lot of tea: black, green, and different kinds of herbal tea, which also are rich in polyphenols.

I'm cutting back on	because	and instead you can
Juice—whether from the grocery store, from juice bars, or home squeezed	High fructose content, along with low fiber, raises blood sugar, which drives inflammation. This is true even for freshly squeezed juice.	Eat whole fruit along with nuts or boiled eggs to stabilize the insulin response. If I buy juice from a juice bar, I like to choose a completely "green" one with low sugar content, that is, made of spinach, ginger, celery, etc.
Sodas	Sugar.	Drink mineral water or "pimped up" water, which I like to call healing water, where I add anti-inflammatory ingredients like green tea, herbs, lime, cucumber, and berries, and let it steep for a while. Another alternative is kombucha, a kind of probiotic "soda" made of fermented green tea that you can find nowadays in many different flavors. Try the one with turmeric and ginger for a maximum boost.
Alcoholic drinks	Alcohol is basically fermented high-sugar plants, and drinking wine, hard liquor, champagne, and other alcoholic drinks affects carbohydrate consumption. As a general guideline, you should only drink alcohol two days a week, since women who drink every day have an increased risk of breast cancer, among other things. A very moderate intake, on the other hand, is linked to anti-inflammation thanks to the polyphenol resveratrol.	Drink moderately—not more than a few glasses a week for someone who wants to live by the book—and think about which drink is the best choice. The anti-inflammatory best practice is red wine with a rich content of the polyphenol resveratrol, which is found in pinot noirs, for example. Certain types of alcohol have a somewhat lower GI value, like dry champagne and vodka. One thing I've learned is to save my drinking for those situations that mean something to me personally and privately and avoid drinking "throwaway" drinks, like wine on airplanes and in more casual work contexts. I like to mix "Rita cocktails," which consist of just a little wine mixed with sparkling mineral water. Or wine mixed with kombucha, which is also a good combination.

THANK YOU
FROM THE AUTHOR

No human being is an island, least of all an author. Writing a book is in one sense solitary work, but I stand on the shoulders of fellow human beings who have supported me with their actions, warmth, and intelligence. I would particularly like to thank the following people:

My publisher, Carina Nunstedt, who helped me see with new eyes and guided me to rethink this journey and dare to write about it in a deeper and more personal way. It's a privilege to be able to work with someone with such intelligence and such a bright light. Thank you also to PR entrepreneur Christina Saliba, who brought us together with her generosity and eye for people.

To my warm and brilliant agent, Rita Karlsson, Kontext Agency, who balances fine-tuned judgment with a true passion for science and quality books.

The fine team at HarperNordic, who radiate such entrepreneurship, such sparkling energy and kindness. Lisa Sorgenfrei, editor; Charlotta Paulson, designer; Sarah Wallskog Lindvall and Annika Widholm, PR and Marketing; Peter Hafverkorn and Pauline Riccius, Sales; as well as Lina Moren, digital director, and managing director Anette Ekström, who understood right away.

In the United States I have been supported by the passionate and professional team at Harper Design, with Christine Choe, Emily Van Derwerken, Dani Segelbaum, Sue Livingston, and uber-cool Marta Schooler. Also in the video room, Marisa Benedetto and Scooter McCray.

The team by the river, HarperCollins UK: thanks for the sterling job and making me feel very welcome to Rachel Kenny, editorial director, Lisa Milton, executive publisher, and Kate Fox, Celia Lomas, Louise McGrory, Georgina Green, Lucy Richardson, and Jennifer Porter.

Rita Catolino—you are a sculptor and people leader, and I'm a different person after being in your circle. To my exercise friends I met through Rita, hugs and thank you, and also to Rita's mother, the fantastic Gabriella.

Annie Wegelius—who with your great creative talent thought of the book's title, and who has lovingly watched over me in so many ways. So many lines of thought in this book have their origin in your intelligence and humanity, and our discussions over the last thirty years.

Then on to my scientific advisors. In Sweden: Professor Inger Björck and the researchers Juscelino Tovar and Anne Nilsson at Lund University—you put me on the right track and have since helped me with my complex material. You deserve all the support in the world for your fantastic work. A warm thank-you as well to Professor Tomas Ekström at Karolinska Institutet, who has provided incredible support by generously reading and giving his opinion about everything. Also at the Karolinska Institutet—thank you to the fantastic Martin Schalling, professor of Neurogenomics, for his thoughts about the accelerated aging that's seen in the case of kidney disease, as well as the brilliant Martin Ingvar, professor of Neurology, for his thoughts on cognition and inflammation.

A warm thank-you also to the wonderful Dr. Anna Marie Olsen, dermatologist at the Dr. Sebagh Clinic in London, and the cutting-edge thinker nutrition physiologist Stephanie Moore at Grayshott Spa in Surrey. In Great Britain I also found Dr. Jeya Prakash, a doctor on Harley Street who is an absolute pioneer whose work points to the future.

In Canada, a big thank-you to my source of inspiration, Dr. Jennifer Stellar at the University of Toronto, who is doing so much original research that I'm in awe. In Denmark, thank you to the incredibly cool professor Bente Klarlund Pedersen at Rigshospitalet in Copenhagen, with her passion for ordinary people's health, which I love; and in India, to Someatheeram Ayurvedic Spa, with its wise and fantastic doctors and therapists.

In the United States, a warm thank-you to the kind Dr. Gary Fraser and his coworkers and to the Information Department at the Loma Linda University.

I also want to thank all the organizers, all the staff, and all the participants at the Bliss camp in California, the Ayurveda course in Kerala, at Grayshott in Great Britain, and at the training camp with Rita Catolino. You have all contributed to my insights, and if any of you in any way feel wrongly described, I ask for your understanding and tolerance, since I've done my best to portray everything in a true and loving way.

A warm thank-you also to my trio of critical readers: my anti-inflammatory friend Jessica Cappelen; my editor at DI, Anna Ekström; as well as my exercise friend Ulrika Fors Stenmarck—you three have been fantastic; you've read, asked questions, given me self-confidence, and shared your opinions.

To Arriane Alexander, who has lived with the book and opened new doors inside me—thank you for your wisdom and your light.

Thanks also to my hairstylist, Lisa Daly, who not only supplies me with vitamins but also is a person I'm in constant dialogue with about life, health, exercise, and food.

To my pod colleague Katrin Marcal, who constantly encouraged me to take new steps and who has been a valuable sounding board when I've had problems.

And to my colleague Karin O'Connor, for brilliance and genuineness in everything, including your book advice.

To the inspiring Jane Sowerby, from Sowerby Housestyle, who assisted with the photography with her innovative and beautiful image styling and wonderful staging. To Jenny Lewis, our super beautiful London photographer, whose photos exude a feeling of light and humanity. And to Pamela and Andrea Makeup, who helped us capture what the book represents.

Thank you also to Mille Broome, for good advice, and to my wonderful meditation sister Annika Dopping and my old dear friend Leni, for being in my Self Help Group, which once started the journey.

Then to my family, who have become involved in the greater journey and the book itself. Greger, you lift and carry me, even when I least deserve it, and walk by my side with such light. Gustaf, you share my interest for exercise, food, and science—thank you for helping with the book's research, which you handled with clever and driving ingenuity and independence when you found several of the leads yourself. Erica, you have been such a professional, smart, and warm reader, always lending an ear to book issues. Jakob, you have filled my heart and my thoughts about inner motivation and goal setting and carried me through some tough phases in the creation of this project. Bisse, thank you for your constant loving encouragement, and for getting me to look at my manuscript with more of a dramatist's eyes. So to all my family—my cup of love overfloweth . . .

REFERENCES

Scientific Interviews

Inger Björck, Professor of Nutritional Science and Director of the Center for Preventive Food Research, Lund University, Lund, Sweden. (Interviews, April–October 2013.)

Tomas Ekström, Professor of Molecular Cell Biology at the Department of Clinical Neuroscience, Karolinska Institutet, Stockholm, Sweden. (Interview and email correspondence from spring 2017 and on.)

Gary Fraser, Cardiologist and Principal Investigator at the Adventist Health Study Project, Loma Linda Medical Center, Loma Linda, California, USA. (Interview, March 2017.)

Bente Klarlund Pedersen, Professor of Integrative Medicine and the Director and founder of Denmark's Research Center for Inflammation and Metabolism, Rigshospitalet, Copenhagen, Denmark. (Interview, May 2017.)

Stephanie Moore, Clinical Nutrition Therapist and Psychologist, as well as Director of Grayshott Spa, Haslemere, Great Britain. (Interview, February 2017.)

Anne Nilsson, Lecturer at the Department of Food Technology, Lund University, Lund, Sweden. (Interview, January 2017.)

Anna Marie Olsen, Clinical Dermatologist, Dr. Sebagh Clinic, London, Great Britain. (Interview, May 2017.)

Jeya Prakash, Physician and specialist on aging at The Medical Park Team, London, Great Britain, and in Chennai, India. (Interview and email correspondence from September 2017 and on.)

Martin Schalling, Professor of Medical Technology at the Department of Molecular Medicine and Surgery, Karolinska Institutet, Stockholm, Sweden. Director at Psykiatrifonden. (Interviews from 2013 on.)

Seena Rajendran, Senior Medical Officer, Somatheeram Ayurveda Resort, Kerala, India. (Interview, January 2016.)

Barry Sears, Director, Inflammation Research Foundation, Peabody, Massachusetts, USA. (Phone interview and email correspondence, spring 2017.)

David Sinclair, Professor, Department of Genetics, Harvard Medical School, Boston, USA. (Email correspondence, spring 2017.)

Jennifer Stellar, Psychologist, Department of Psychology, University of Toronto, Canada. (Interview, January 2017.)

Juscelino Tovar, Lecturer and Research Project Director, Center of Preventive Nutrition Research, Lund University, Lund, Sweden. (Interview, January 2017.)

In addition, during my thirty years as a science journalist, I've talked to a large number of researchers in disciplines like immunology, nutrition, genetics, endocrinology, sports physiology, and psychiatry, not to mention all the people I have met during my many trips to India who have told me about Ayurveda. Insight is gained step by step, and all my conversations with the fitness specialist Rita Catolino have also meant a great deal. The responsibility for the final presentation of all the puzzle pieces is entirely my own, however.

Scientific Articles

James David Adams et al. (2012). Mugwort (Artemisia vulgaris, Artemisia douglasiana, Artemisia argyi) in the treatment of menopause, premenstrual syndrome, dysmenorrhea and attention deficit hyperactivity disorder. *Chinese Medicine*, vol. 3, no. 3/2012, pp. 116–123.

Joanne S. Allard et al. (2009). Dietary activators of Sirt1. *Molecular Cell Endocrinology*, vol. 299, no. 1/2009, pp. 58–63.

Jessica A. Alvarez et al. (2009). Fasting and postprandial markers of inflammation in lean and overweight children. *American Journal of Clinical Nutrition*, vol. 89, no. 4/2009, pp. 1138–1144.

Anthony T. Annunziato (2008). DNA packaging: Nucleosomes and chromatin. *Nature Education*, vol. 1, no. 1/2008, p. 26.

John Axelsson et al. (2013). Effects of sustained sleep restriction on mitogen-stimulated cytokines, chemokines and T helper 1/ T helper 2 balance in humans. *PloS One*, vol. 8, no. 12/2013.

Yang Bai et al. (2017). Awe, the diminished self, and collective engagement: Universals and cultural variations of self. *Journal of Personality and Social Psychology*, vol. 113, no. 2/2017, pp. 185–209.

Nir Barzilai & Ilan Gabrieli (2010). Genetic studies reveal the role of the endocrine and metabolic systems in aging. *The Journal of Clinical Endocrinology and Metabolism*, vol. 95, no. 10/2010, pp. 4493–4500.

Jessica E. Beilharz et al. (2016). Short-term exposure to a diet high in fat and sugar, or liquid sugar, selectively impairs hippocampal-dependent memory, with differential impacts on inflammation. *Behavioural Brain Research*, vol. 306/2016, pp. 1–7.

Jessica E. Beilharz et al. (2016). The effect of short-term exposure to energy-matched diets enriched in fat or sugar on memory, gut microbiota and markers of brain inflammation and plasticity. *Brain, Behavior, and Immunity*, vol. 57/2016, pp. 304–313.

Alessandra Bordoni et al. (2015). Dairy products and inflammation: A review of the clinical evidence. Critical Reviews in *Food Science and Nutrition*, vol. 57, no. 12/2015, pp. 2497–2525.

William M. Brown (2015). Exercise-associated DNA methylation change in skeletal muscle and the importance of imprinted genes: a bioinformatics meta-analysis. *British Journal of Sports Medicine*, vol. 49, no. 24/2015, pp. 1567–1578.

Ewa Bulzacka et al. (2016). Chronic peripheral inflammation is associated with cognitive impairment in schizophrenia: Results from the multicentric FACE-SZ dataset. *Schizophrenia Bulletin*, vol.42, no. 5/2016, pp. 1290–1302.

Ivana Celic et al. (2006). The sirtuins Hst3 and Hst4p preserve genome integrity by controlling histone H3 lysine 56 deacetylation. *Current Biology*, vol. 16, no. 13/2006, pp. 1280–1289.

Jennifer Couzin-Frankel (2011). Aging Genes: the sirtuin story unravels. *Science*, vol. 334, no. 6060/2011, pp. 1194–1198.

David Creswell et al. (2016). Alterations in resting-state functional connectivity link mindfulness meditation with reduced interleukin-6: A randomized controlled trial. *Biological Psychiatry*, vol. 80, no. 1/2016, pp. 53–61.

Weiwei Dang et al. (2009). Histone H4 lysine 16 acetylation regulates cellular lifespan. *Nature*, no. 459, pp. 802–807.

J. Denham et al. (2014). Exercise: putting action into our epigenome. *Sports Medicine*, vol. 44, no. 2/2014, pp. 189–209.

Sally Dickerson et al. (2004). Immunological Effects of Induced Shame and Guilt. *Psychosomatic Medicine*, vol. 66, no. 1/2004, pp. 124–131.

J.A. Dumas et al. (2016). Dietary saturated fat and monounsaturated fat have reversible effects on brain function and the secretion of pro-inflammatory cytokines in young women. *Metabolism*, vol. 65, no. 10/2016, pp. 1582–1588.

Tobias Ehlert, Perikles Simon, & Dirk A. Moser (2013). Epigenetics in sports. *Sports Medicine*, vol. 43, no. 2/2013, pp. 93–110.

Sophie Erhardt et al. (2001). Kynurenic acid levels are elevated in cerebrospinal fluid of patients with schizophrenia. *Neuroscience Letters*, vol. 313, no. 1–2/2001, pp. 96–98.

Marjo H. Eskelinen & Miia Kivipelto (2010). Caffeine as a protective factor in dementia and Alzheimer's disease. *Journal of Alzheimer's Disease*, vol. 20, suppl. 1-2010, pp. 167–174.

M. F. Facheris et al. (2008). Coffee, caffeine-related genes, and Parkinson's disease: A case-controlled study. *Movement Disorders*, vol. 23, no. 14/2008, pp. 2033–2040.

Elinor Fondell et al. (2011). Short natural sleep is associated with higher T cell and lower NK cell activities. *Brain, Behavior, and Immunity*, vol. 25, no. 7/2011, pp. 1367–1375.

Yuanquin Gao et al. (2017). Dietary sugars, not lipids, drive hypothalamic inflammation. *Molecular Metabolism*, vol. 6, no. 8/2017, pp. 897–908.

Amie M. Gordon et al. (2017). The dark side of the sublime: Distinguishing a threat-based variant of awe. *Journal of Personality and Social Psychology*, vol. 113, no. 2/2017, pp. 310–328.

Jean-Philippe Gouin et al. (2012). Stress, negative emotions, and inflammation. I: *The Oxford Handbook of Social Neuroscience*, Oxford University Press. http://www.oxfordhandbooks.com /view/10.1093/oxfordhb/9780195342161.001.0001/oxfordhb-9780195342161-e-054. Visited November 27, 2017.

Mai-Lis Hellénius (2011). Metabola syndromet [pdf]. FYSS—fysisk aktivitet i sjukdomsprevention och sjukdomsbehandling. http://fyss.se/wp-content/uploads/2011/02/32.-Metabola-syndromet.pdf. Downloaded November 16, 2017.

Edel Hennessy et al. (2016). Systemic TNF—produces acute cognitive dysfunction and exaggerated sickness behavior when superimposed upon progressive neurodegeneration. *Brain, Behavior, and Immunity*, vol. 59/2017, pp. 233–244.

Nolan J. Hoffman et al. (2015). Global phosphoproteomic analysis of human skeletal muscle reveals a network of exercise-regulated kinases and AMPK substrates. *Cell Metabolism*, vol. 22, no. 5/2015, pp. 922–935.

Shin-ichiro Imaj & Leonard Guarente (2016). It takes two to tango: NAD+ and sirtuins in aging/ longevity control. *Npj Aging and Mechanisms of Disease*, vol. 2/2016.

S. Intahphuak, P. Khonsung, & A. Panthong (2009). Anti-inflammatory, analgesic, and antipyretic activities of virgin coconut oil. *Pharmaceutical Biology*, vol 48, no. 2/2010, pp. 151–157.

Shorena Janelidze et al. (2011). Cytokine levels in the blood may distinguish suicide attempters from depressed patients, *Brain, Behaviour, and Immunity*, vol. 25, no. 2/2011, pp. 335–339.

Nancy S. Jenny (2012). Inflammation in aging: cause, effect, or both? *Discovery Medicine*, vol. 13, no. 73/2012, pp. 451–460.

Neha John-Henderson et al. (2015). Socioeconomic status and social support: Social support reduces inflammatory reactivity for individuals whose early-life socioeconomic status was low. *Psychological Science*, vol. 26, no. 10/2015, pp. 1620–1629.

James Joseph et al. (2009). Nutrition, brain aging, and neurodegeneration. *The Journal of Neuroscience*, vol. 29, no. 41/2009, pp. 12795–12801.

Jian X. Kang & Karsten H. Weylandt (2008). Modulation of inflammatory cytokines by omega-3 fatty acids. *Sub-cellular Biochemistry Book Series, SBI*, vol. 49/2008, pp. 133–143.

Janice K. Kiecolt-Glaser et al. (2002). Emotions, morbidity, and mortality: new perspectives from psychoneurology. *Annual Review of Psychology*, vol. 53/2002, pp. 83–107.

Janice K. Kiecolt-Glaser et al. (2002). Psychoneuroimmunology: psychological influences on immune function and health. *Journal of Consulting and Clinical Psychology*, vol. 70, no. 3/2002, pp. 537–547.

Tania S. King-Himmelreich et al. (2016). The impact of endurance exercise on global and AMPK gene-specific DNA methylation. *Biochemical and Biophysical Research Communications*, vol. 474, no. 2/2016, pp. 284–290.

Bente Klarlund Pedersen (2011). Muscles and their myokines, *Journal of Experimental Biology*, vol. 214, no. 2/ 2011, pp. 337–346.

Harold Koenig et al. (1998). The relationship between religious activites and blood pressure in older adults. *The International Journal of Psychiatry in Medicine*, vol. 28, no. 2/1998, pp. 189–213.

Harold Koenig (2004). Religion, spirituality, and medicine: Research findings and implications for clinical practice. *Southern Medical Journal*, vol. 97, no. 12/2004, pp. 1194–1200.

J. P. Kooman et al. (2014). Chronic kidney disease and premature ageing. Nature Reviews. *Nephrology*, vol. 10, no. 12/2014, pp. 732–742.

Marie-Ève Labonté et al. (2013). Impact of dairy products on biomarkers of inflammation: a systematic review of randomized controlled nutritional intervention studies in overweight and obese adults. *American Journal of Clinical Nutrition*, vol. 97, no. 4/2013, pp. 706–717.

Heidi Ledford (2011). Longevity genes challenged: Do sirtuins really lengthen lifespan? *Nature*, September 21, 2011. (http://www.nature. com/news/2011/110921/full/news.2011.549.html) Visited November 15, 2017.

Shanshan Li et al. (2016). Association of religious service attendance with mortality among women. *JAMA Internal Medicine*, vol. 176, no. 6/2016, pp. 777–785.

Dorothy Long Parma et al. (2015). Effects of six months of yoga on inflammatory serum markers prognostic of recurrence risk in breast cancer survivors. *Springer Plus*, vol. 4, no. 143/2015.

E. Lopez-Garcia et al. (2005). Consumption of trans fatty acids is related to plasma biomarkers of inflammation and endothelial dysfunction. *The Journal of Nutrition*, vol. 135, no. 3/2005, pp. 562–566.

Karin Luttropp et al. (2009). Genetics/Genomics in chronic kidney disease—towards personal-ized medicine? *Seminars in Dialysis*, vol. 22, no. 4/2009, pp. 417–422.

Arndt Manzel et al. (2014). Role of "western diet" in inflammatory autoimmune diseases. *Current Allergy and Asthma Reports*, vol. 14, no. 1/2014, pp. 404.

Evi M. Merchen et al. (2013). Calorie restriction in humans inhibits the PI3K/AKT pathway and induces a younger transcription profile. *Aging Cell*, vol. 12, no. 4/2013, pp. 645–651.

Esmaeil Mortaz et al. (2015). Anti-inflammatory effects of Lactobacillus rahmnosus and Bifidobacterium breve on cigarette smoke activated human macrophages. *PloS One*, vol. 10, no. 8/2015.

Raul Mostoslavsky et al. (2010). At the crossroad of lifespan, calorie restriction, chromatin and disease: Meeting on sirtuins. *Cell Cycle*, vol. 9, no. 10/2010, pp. 1907–1912.

Dariush Mozaffarian et al. (2004). Dietary intake of trans fatty acids and systemic inflammation in women. *American Journal of Clinical Nutrition*, vol. 79, no. 4/2004, pp. 606–612.

Janet M. Mullington et al. (2013). Sleep loss and inflammation. *Best Practice & Research: Clinical Endocrinology & Metabolism*, vol. 24, no. 5/2010, pp. 775–784.

Jens P. Nilsson et al. (2015). Less effective executive functioning after one night's sleep deprivation. *Journal of Sleep Research*, vol. 14, no. 1/2005, pp. 1–6.

Tiago Fleming Outeiro et al. (2007). Sirtuin 2 inhibitors rescue α-Synuclein-mediated toxicity in models of Parkinson's disease. *Science*, vol. 317, no. 5837/2017, pp. 516–519.

Helios Pareja-Galeano et al. (2014). Physical exercise and epigenetic modulation: elucidating intricate mechanisms. *Sports Medicine*, vol. 44, no. 4/2014, pp. 429–436.

Roberto Pecoits-Filho et al. (2003). Genetic approaches in the clinical investigation of complex disorders: malnutrition, inflammation, and atherosclerosis (MIA) as a prototype. *Kidney International*, vol. 63, suppl. 84/2003, pp. 162–167.

Karin de Punder & Leo Pruimboom (2013). The dietary intake of wheat and cereal grains and their role in inflammation. *Nutrients*, vol. 5, no. 3/2013, pp. 771–787.

J. Schemies et al. (2010). NAD^+-dependent histone deacetylases (sirtuins) as novel therapeutic targets. *Medicinal Research Reviews*, vol. 30, no. 6/2010, pp. 861–889.

Barry Sears (2015). Anti-inflammatory diets. *Journal of the American College of Nutrition*, vol. 34, suppl. 1/2015, pp. 14–21.

Barry Sears (2016). Delaying adverse health consequences of aging: the role of omega 3 fatty acids on inflammation and resoleomics. *CellR4*, vol. 4, no. 4/2016.

Barry Sears & Camillo Ricordi (2012). Role of fatty acids and polyphenols in inflammatory gene transcription and their impact on obesity, metabolic syndrome and diabetes. *European Review for Medical and Pharmacological Sciences*, vol. 16, no. 9/2012, pp. 1137–1154.

Paul G. Shields et al. (2017). The role of epigenetics in renal aging. Nature Reviews. *Nephrology*, vol. 13, no. 8/2017, pp. 471–482.

Guido Shoba et al. (1997). Influence of piperine on the pharmacokinetics of curcumin in animals and human volunteers. *Planta Medica*, vol. 64, no. 4/1998, pp. 353–356.

Fabíola Lacerda Pires Soares et al. (2013). Gluten-free diet reduces adiposity, inflammation and insulin resistance associated with the induction of PPAR-alpha and PPAR-gamma expression. *The Journal of Nutritional Biochemistry*, vol. 24, no. 6/2013, pp. 1105–1011.

Jennifer Stellar et al. (2015). Positive affect and markers of inflammation: Discrete positive emotions predict lower levels of inflammatory cytokines. *Emotion*, vol. 15, no. 2/2015, pp. 129–133.

Jennifer Stellar et al. (2017). Awe and Humility. *Journal of Personal Social Psychology*, August 31, 2017.

Peter Stenvinkel et al. (2007). Impact of inflammation on epigenetic DNA methylation—a novel risk factor for cardiovascular disease? *Journal of Internal Medicine*, vol. 261, no. 5/2007, pp. 488–499.

Ambarish Vijayaraghava et al. (2015). Effect of yoga practice on levels of inflammatory markers after moderate and strenuous exercise, *Journal of Clinical and Diagnostic Research*, vol. 9, no. 6/2015, pp. CC08–CC12.

Nannan Zhang et al. (2016). Calorie restriction-induced SIRT6 activation delays aging by supressing NF-kB signaling. *Cell Cycle*, vol. 15, no. 7/2016, pp. 1009–1018.

News Articles

Sarah Knapton (2017). Depression is a physical illness which could be treated with anti-inflammatory drugs, scientists suggest. *The Daily Telegraph*, 8 September 2017.

Thomas Lerner (2015). Svenskar tror—men inte på Gud. *Dagens Nyheter*, 26 May 2015.

Books in Which I've Found Facts and Inspiration

Dan Buettner (2009). *The blue zones: Lessons for living longer from the people who've lived the longest*. National Geographic Society.

Floyd H. Chilton (2009). *The Gene-smart diet: The revolutionary eating plan that will rewrite your genetic destiny—and melt away the pounds*. Rodale Books.

Vicky Edgson & Adam Palmer (2015). *Gut gastronomy: Revolutionise your eating to create great health*. Jaqui Small LLP.

Yuval Noah Harari (2017). *Homo deus: En kort historik över morgondagen*. Natur & Kultur.

Donald C. Johanson & Maitland A. Edey (1981). *Lucy: The beginnings of humankind*. Simon & Schuster.

Bente Klarlund Pedersen (2010). *Sandheden om sundhed*. Politikens forlag.

Nigella Lawson (2015). *Simply Nigella: Feel good food*. Random House UK.

Sunil Pai (2016). *An inflammation nation: The definitive 10-step guide to preventing and treating all diseases through diet, lifestyle, and the use of natural anti-inflammatories*. Rocdoc Publications.

Nicholas Perricone (2011). *Forever young: The science of nutrigenomics for glowing, wrinkle-free skin and radiant health at every age*. Simon & Schuster.

Tosca Reno (2010). *Your best body now: Look and feel fabulous at any age the eat-clean way*. Harlequin.

Robert M. Sapolsky (2003). *Varför zebror inte får magsår*. Natur & Kultur.

David Servant-Schreiber (2011). *Anticancer—ett nytt sätt att leva*. Natur & Kultur.

Janesh Vaidya & Malin Barrling (2013). *Maten är min medicin: Ayurveda i ditt kök*. Norstedts.

Sites with Facts and Inspiration

Bluezones.com

This site summarizes the extremely interesting book about the Blue Zones written by Dan Buettner, who describes, with warmth and in medical detail, how the research team went about studying these zones. Here you can also read about how the people there live lives filled with purpose, peace, vegetables, and good bacteria.

Sirtfooddiet.net

Sirtfood diet is a diet based on sirtuins, the epigenetic regulators that decrease inflammation, and it's based on eating food that is rich in polyphenols, among other things. The site explores the anti-inflammatory lifestyle and contains both facts and recipes.

INDEX OF RECIPES